LANDSCAPE, TOURISM, AND MEANING

New Directions in Tourism Analysis

Series Editor: Dimitri Ioannides, Missouri State University, USA

Although tourism is becoming increasingly popular as both a taught subject and an area for empirical investigation, the theoretical underpinnings of many approaches have tended to be eclectic and somewhat underdeveloped. However, recent developments indicate that the field of tourism studies is beginning to develop in a more theoretically informed manner, but this has not yet been matched by current publications.

The aim of this series is to fill this gap with high quality monographs or edited collections that seek to develop tourism analysis at both theoretical and substantive levels using approaches which are broadly derived from allied social science disciplines such as Sociology, Social Anthropology, Human and Social Geography, and Cultural Studies. As tourism studies covers a wide range of activities and sub fields, certain areas such as Hospitality Management and Business, which are already well provided for, would be excluded. The series will therefore fill a gap in the current overall pattern of publication.

Suggested themes to be covered by the series, either singly or in combination, include – consumption; cultural change; development; gender; globalisation; political economy; social theory; sustainability.

Also in the series

Tourism and the Branded City
Film and Identity on the Pacific Rim
Stephanie Hemelryk Donald and John G. Gammack
ISBN 978-0-7546-4829-1

Raj Rhapsodies: Tourism, Heritage and the Seduction of History
Edited by Carol E. Henderson and Maxine Weisgrau
ISBN 978-0-7546-7067-4

Tourism and Borders
Contemporary Issues, Policies and International Research
Edited by Helmut Wachowiak
ISBN 978-0-7546-4775-1

Christian Tourism to the Holy Land
Pilgrimage during Security Crisis
Noga Collins-Kreiner, Nurit Kliot, Yoel Mansfeld and Keren Sagi
ISBN 978-0-7546-4703-4

Landscape, Tourism, and Meaning

Edited by

DANIEL C. KNUDSEN
MICHELLE M. METRO-ROLAND
ANNE K. SOPER
CHARLES E. GREER
Indiana University, Bloomington, USA

LONDON AND NEW YORK

First published 2008 by Ashgate Publishing

2 Park Square, Milton Park, Abingdon, Oxon OX14 4RN
711 Third Avenue, New York, NY 10017, USA

Routledge is an imprint of the Taylor & Francis Group, an informa business

First issued in paperback 2016

British Library Cataloguing in Publication Data
Landscape, tourism, and meaning. - (New directions in
    tourism analysis)
    1. Tourism 2. Tourism - Social aspects 3. Culture and
    tourism 4. Landscape assessment
    I. Knudsen, Daniel C., 1955-
    306.4'819

Library of Congress Cataloging-in-Publication Data
Landscape, tourism, and meaning / by Daniel C. Knudsen ... [et al.].
    p. cm. -- (New directions in tourism analysis)
    Includes bibliographical references and index.
    ISBN 978-0-7546-4943-4 (alk. paper)
    1. Tourism. 2. Tourism--Social aspects. 3. Culture and tourism. 4. Landscape assessment.
I. Knudsen, Daniel C., 1955-

    G155.A1L284 2008
    306.4'819--dc22

                                                                              2007046439

ISBN 13: 978-0-7546-4943-4 (hbk)
ISBN 13: 978-1-138-25528-9 (pbk)

Contents

"Other"   43
Conclusion   48

5   **Mauritian Landscapes of Culture, Identity, and Tourism**   **51**
       Anne K. Soper
       Introduction   51
       Methods   51
       The Making of Mauritius   52
       Mauritian Identity   52
       Identity and Tourism   54
       Developing Mauritian Cultural Heritage Tourism   56
       Examples from the Mauritius Tourism Landscape   58
       Conclusion   63

6   **Slicing the *Dobish Torte*: The Three Layers of Tourism in Munich** **65**
       Richard Wolfel
       Introduction   65
       Lefebvre and the Production of Space   66
       Tourism and Landscapes   67
       The Three Layers of Tourism and the Production of Space in
           Post-World War II Munich   68
       Conclusion   73

7   **A Nostalgia for Terror**   **75**
       Michelle M. Metro-Roland
       Introduction   75
       Locating   76
       Looking   81
       Conclusion   93

8   **The Parallax of Landscape: Situating Celaque
       National Park, Honduras**   **95**
       Benjamin F. Timms
       Introduction   95
       National Parks as Landscape   95
       The Morphology of Celaque National Park   98
       The Insider versus Outsider Landscape Perspectives   101
       The Landscape of Social Formation   102
       Organic versus Universal Landscape   104
       Landscape as Social Compromise   105
       Conclusion   106

9   **Insiders and Outsiders in Thy**   **109**
       Daniel C. Knudsen
       Introduction   109
       Insider versus Outsider   110

List of Figures and Maps

List of Contributors

Lisa C. Braverman is a researcher and editor living in Bloomington, Indiana. Her research interests include cultural travel, the relationship of study abroad experiences to performance and rhetoric, and the potential benefits and pitfalls of international education. Braverman has spent significant time in Denmark, Poland and Great Britain, which prompted her interest in the deconstruction of overseas experiences.

Shanon Donnelly is a graduate student in the Department of Geography and a lecturer in the Department of Geography at University of Akron in Ohio. The interdisciplinary nature of this training has cultivated his research interests in how the spatial organization of land tenure structures the aggregation of individual land management decisions in forested environments. From landscapes dominated by small individually-owned parcels in American exurbs to large communal holdings in southern Mexico, understanding the emergent outcomes of complex spatial interaction in the landscape is a theme that underlies his various research endeavors.

Yamir González-Vélez attended the University of Puerto Rico, Rio Piedras Campus, from 1995 to 2000. She earned a BA degree in secondary education with a concentration in social science, as well as a second major in Geography. During her undergraduate years, she was a McNair Scholar, an intern for the USDS Forest service, a research assistant for the Naturation Research Program, and an intern for the NOAA. Currently, she is working on a dual MA in Geography and Latin American and Caribbean Studies at Indiana University Bloomington.

Charles Greer is an associate professor (emeritus) of Geography and East Asian Languages and Cultures at Indiana University. His research interests include conservation, natural resources, and landscape studies. He conducts field research in Indiana, the American West, China, and Mexico. He earned his Ph.D. in Geography at the University of Washington in 1975. In addition to his academic writing, he is a published poet.

Sean Huff is with the U.S. Agency for International Development Foreign Service, working as a USAID/Russia Project Development Officer. His research interests are in urban landscape and urban planning. Mr. Huff received his Masters degree from Indiana University in 2002. Prior to coming to Indiana University, he served with the Peace Corps in Nepal.

Daniel C. Knudsen is a professor in the Department of Geography and director of the International Studies program at Indiana University. His research interests include cultural economy, rural geography, landscape, and tourism. Professor Knudsen earned a Ph.D. in Geography with a minor in Economics from Indiana

University in 1984. He was a Fulbright Fellow in Denmark in 1995 and returns there annually to pursue field research.

Michelle M. Metro-Roland is a doctoral candidate in the Department of Geography at Indiana University. Her research is situated at the interstices of landscape and tourism. She has worked across geographic scales from small rural sites in the Midwest of the United States to the urban built environments of European capital cities such as Skopje and Budapest. She received a Fulbright-Hayes award for support of her dissertation research and spent the 2005–2006 academic year in Budapest, Hungary investigating the saliency of Hungarian cultural identity within the prosaic spaces of the inner city.

Jillian M. Rickly obtained Bachelor of Arts degrees in Biology and Geography from Indiana University in 2005. In the spring of 2006, she began work in the Master of Arts program in the Department of Geography at Indiana University. Her current research interests are in landscape studies with particular interest in symbolic landscapes and power representations, as well as the cultural ecology and environmental perceptions of indigenous populations.

Anne K. Soper is an assistant professor in the Tourism and Events Management Program at George Mason University Ras Al Khaimah. Her work focuses on cultural heritage tourism in less developed island nations, tourism planning and development, and issues of sustainable tourism. She has spent most of her life engaged in travel and tourism. Professor Soper earned a Ph.D. in Geography from Indiana University in 2006.

Benjamin F. Timms earned his B.A. in Geography at the University of New Mexico and M.A. and Ph.D. in Geography at Indiana University. While his initial studies and research were in neo-Marxist political economy, under the tutelage of Dr. Charles Greer and Dr. Dan Knudsen he developed a keen interest in the humanistic power of landscape. His current research broadly focuses on human-environment interaction in the Caribbean and Central America, with more specific interests in Political Ecology, Landscape, and Development. Benjamin Timms currently serves as an assistant professor of Geography at California Polytechnic Institute.

Richard Wolfel is an assistant professor in the Department of Geography and Environmental Engineering at the United States Military Academy. His specialties include the influence of nationalism on urban development, migration in transitional societies, rural poverty and political aspects of globalization. Professor Wolfel earned a Ph.D. in Geography, with a minor in Central Eurasian Studies, from Indiana University in 2001. He has written numerous articles for publications such as *Nationalities Papers* and *Journal of Central Asian Studies*.

Altynai Yespembetova is a lecturer at the Kazakhstan Institute of Management, Economics and Strategic Research. Her specialties include tourism and economic development in Central Asia. Ms. Yespembetova received her Masters degree from

Indiana University in 2005. She is currently pursuing her Ph.D. at Kazakh National University in Almaty.

Preface

This book presents papers which originated from two specially-organized sessions of the annual meeting of the Association of American Geographers held in Philadelphia, Pennsylvania, in March 2004. The sessions themselves emerged from a growing dialogue between specialists in landscape studies and those in tourism at Indiana University in 2003–2004, both of whom were grappling with similar issues of presentation, interpretation and meaning.

The chapters break fresh ground by linking the emerging field of tourism theory to developments in the established field of landscape studies. Issues of identity are a common thread throughout and are raised with regard to the social construction of landscape and its portrayal through tourism. The chapters range in scale from regional to national, personal to political, and from local residents to international tourists. The multiplicity of interpretations and meanings between these scales are focal points within each chapter. Case studies from across the world are used to demonstrate the extent to which landscape theory and tourism practice come together within the realm of various countries, regions, and cities. Although the chapters are all written by geographers, the topics of both landscape and tourism are multidisciplinary in nature and will appeal to a wide variety of readers in the social sciences and humanities.

Daniel C. Knudsen, Michelle M. Metro-Roland, Anne K. Soper, and Charles E. Greer

Acknowledgements

There are many who need to be recognized for their contributions to making this edited volume possible, but some deserve special mention. First, this volume could not have been possible without the perseverance and keen editing skills of Lisa C. Braverman, who has worked patiently and diligently to see this effort through. Second, the editors wish to thank Val Rose, Elaine Rivers, and Neil Jordan at Ashgate for their encouragement and patience in answering our many questions.

Additionally, the authors as a group would like to thank our students who sat through the lectures wherein we first tried out and later honed many of the ideas contained in this volume and our colleagues who gave us valuable feedback on our ideas. Some colleagues and students have been kind enough to read and comment on earlier versions of paper drafts. Finally, we must thank friends, spouses and, in some cases, children who suffered neglect as we worked on finishing the manuscript.

Dedicated to

Jane T. Zaring

who first explored the nexus of landscape, tourism, and meaning

Chapter 1

Landscape, Tourism, and Meaning: An Introduction

Daniel C. Knudsen, Anne K. Soper, and Michelle M. Metro-Roland

The purpose of this book is to re-theorize tourism. To do this, we draw less on Foucauldian notions of the gaze than on the social construction of meaning in the landscape. We argue that in any view of the landscape and in tourism generally, there is a multiplicity of insider and outsider meanings. By theorizing tourism in this manner, we are situating the study of tourism within the framework of social theory. In grounding tourism studies within the framework of social theory, and particularly in the social theoretic approaches to landscape, we hope to better tease out both the ways in which tourist sites are artfully constructed and the ways that tourism landscapes are filled with intended and unintended meaning for the tourist. At the same time, we also wish to suggest that it is impossible to separate tourism from issues of identity involving both the tourist and the receiving population. This argument stems from landscape's role as both locus of tourism and reification of identity. Thus, just as a place's landscape is the built-up consequences of a place's identity process, so too is tourism the practice of deciphering identity from clues in the landscape of a place.

Background to the Current Theory

The current theorization of tourism has its basis in the early work of Michel Foucault, who is among the most influential of twentieth-century thinkers. Foucault spent his life researching the relationships between power and knowledge. Unlike Francis Bacon, who held that "knowledge is power," Foucault held that "power is knowledge." To Foucault, discourse is the medium through which power is transformed into knowledge and knowledge cannot exist outside of discourse (Foucault 1980). All "known" things have their corresponding discourse that gives them meaning and all discourses have non-discursive elements that may undermine or transform the discourse. According to Foucault, "we must make allowance for the concept's complex and unstable process whereby discourse can be both an instrument and an effect of power, but also a hindrance, a stumbling block, a point of resistance and a starting point for an opposing strategy" (Foucault 1990, 101). In Foucault's mind power has its genesis not in a given discourse, but in the structural arrangements which surround that discourse. These structural arrangements sanction a discourse as legitimate or illegitimate and typically include use of a specialized vocabulary, access to certain facilities, or certain credentials.

A second important aspect of Foucault's work is that, like most twentieth century scholars, he privileges sight or vision as the most important of the human senses. Foucault's assessments of visual practices in certain environments yielded the exceptionally important concept termed "*le regard*," or in English, "the gaze." In *The Birth of the Clinic* (1973), Foucault discusses the power associated with the gaze in a medical setting by presenting the active vision attributed to physicians and how their observations become the discourse which is passed on as truth to students of medicine and patients under treatment:

> Over all these endeavors on the parts of clinical thought to define its methods and scientific norms hovers the great myth of pure Gaze that would be pure Language: a speaking eye. It would scan the entire hospital field, taking in and gathering together each of the singular events that occurred within it; and as it saw, as [it] saw ever more and more clearly, it would be turned into speech that states and teaches ... [1973, 89]

Foucault concludes that hospitals are superficially places providing medical care, but are intrinsically "a sort of semi-juridical structure, an administrative entity which, along with the already constituted powers, and outside the courts, decides, judges, and executes" (Foucault 1984, 125). Ultimately, the doctor as representative of the larger medical institution seems to embody the power, though it emanates from the institution of medical language itself (Foucault 1984).

Foucault's discussion of the connection between the gaze and power relations is further exemplified in his research on the history of prisons as mechanisms for punishment and oppression. Here Foucault examines an architectural structure known as the Panopticon, a well-designed and situated prison consisting of one-man cells and a tower from which prisoners could be visually monitored by a single guard. The guard is granted the all-encompassing gaze while the prisoners are unable to see one another or the guard (Foucault 1978). The power exerted here is different from that of the doctor because the prisoners never really know when the guard is watching. The knowledge that the guard may be watching and waiting to punish rule violations imposes a form of self-monitoring upon the prisoners (Foucault 1984). Foucault remarks:

> There is no need for arms, physical violence, or material constraints – just a gaze. An inspecting gaze, a gaze which each individual under its weight will end by interiorizing to the point where he is his own overseer, each individual thus exercising this surveillance over, and against, himself. [1980, 155]

But whereas the gaze is discursive in *The Birth of the Clinic*, here the physical space acts as a contributing non-discursive component of the effect of the gaze.

Urry and the Tourist Gaze

Current constructs of tourism theory date from the publication of Urry's *The Tourist Gaze* (1990) and his formulation of the concept of tourism as an exercise in

Foucauldian gazing as delineated in *Birth of the Clinic* (Foucault 1973). In Urry's conceptualization, Foucault's doctor is replaced by the tour guide who directs the gaze of the tourists and tells them how to interpret a given sight, while the patients are replaced by the inhabitants and sights of the host country. In this manner of thinking, tourists go forth looking for the unique, distinct, and unusual—somewhere exotic, maybe erotic, but certainly someplace that offers an atmosphere different from that to which they are accustomed (Urry 1992b). Like doctors watching mad men, tourists desire to gaze at places and especially people who are very different from themselves (Urry 1992a). The visual display of people, places, and things makes them forms of "spectacle" and tourism involves the "spectaclization of place" (Urry 1992a, 5). For Urry (1990), the tourist gaze is all consuming and paradoxical. It is all-consuming because it defines the post-modern world. It is paradoxical because tourists both transform and are transformed by their gazing (Urry 1990).

Urry's research is central to the theorization in several ways. First, it redefines the tourism industry so that "the fundamental characteristic of tourist activity is to look upon particular objects or landscapes which are different from the tourist's everyday experiences" (Gaffey 2004, 4). Second, it redefines tourism as an increasingly "signposted experience" involving the "spectaclization of place" (Urry 1995, 139; see also MacCannell 1976 for an earlier treatment) wherein, the "tourist industry" is built around the production of symbols to be gazed upon (Urry 1990, 101). Third, Urry points out that tourism contains numerous face-to-face interactions that must be highly managed and that give rise to particular forms of social relations. Fourth, the tourist gaze creates a situation whereby culture and tourism become one, since culture is increasingly commodified as spectacle for consumption by tourists. Urry uses Lash's term "de-differentiation" to illustrate this concept (1995, 148; 2002, 75). This notion leads to what has been termed the "end of tourism" because in the post-modern world we are engaged in tourism "as continual practice" (Rojek and Urry 1997) and the distinction between representation and reality slips away (Urry 1995).

Tourism researchers have responded to Urry's theorization in a variety of ways. First, Urry's thesis that tourism involves gazing at "particular objects or landscapes" has proven problematic. The term "gazing" cannot be taken literally, since tourism is an experience that involves all of the senses and differing kinds of tourism draw to varying degrees on these other senses (Dann and Jacobsen 2003). This has led scholars to move away from the notion of gazing to one more closely aligned with "performance" (Edensor 2000; Perkins and Thorns 2001). Second, Urry's reframing of tourist sites as "spectacles of place" wherein the object of the gaze is typically an artful construction of signs (Midtgard 2003) avoids any discussion of the varying degrees of authenticity in these artful constructions. Subsequent research indicates that tourists are drawn by spectacle in all its myriad forms—from "natural" through "pastoral" and "heritage" to "fantasy" (Lisle 2004). Third, Urry's theorization of tourism overemphasizes the role played by tourists. In reality, the social relations surrounding tourism are complex and must be negotiated, contested, and resisted (Kirtsoglou and Theodossopoulos 2004). Paradoxically, MacCannell (2001) has critiqued Urry as presenting too structuralist of an argument, and instead he posits the existence of a second gaze which allows for tourist agency in seeking the authentic which lies just behind the surface. Lastly, Urry's tying of tourism to a desire to commodify nature (which in turn stems from a yearning for escape from the banal world of globalized capitalism with its predicable social relations and limited possibilities for self-understanding) does not get us any closer to understanding what compels people to tour.

A Landscape Approach to Tourism Theory

The starting point for a theory of tourism is the notion of seeing or reading the landscape and interpreting its meaning. Tourism should generally be understood as a discourse among three sets of actors: 1) tourists; 2) locals; and 3) intermediaries, including government ministries, travel agents, and tourism promotion boards (Nash 1996). This suggests that if tourism is spectacle, then surely there are multiple parties involved in the creation of this spectacle (Squire 1994).

We would argue that because landscapes must be read, they are open to multiple interpretations—they are heterotopic. This is the case because "our experience of any landscape through the senses is inseparable from the social and psychological context of the experience" (Sopher 1979, 138). However, while personal reaction to events and texts are not privileged, societal institutions and structures do introduce a certain commonality or concordance to personal meaning.

Our thoughts once written or spoken, however, are immediately thrust into contexts characterized by asymmetric power relations where some discourse is privileged while other discourses are disparaged. Meaning and knowledge derived by this process cannot be separated from the process of its creation. That is to say, the meaning of landscape, like all meaning, is created, recreated, and contested in social process (Mitchell 2001). Simply put, objectivity in landscape study does not and cannot exist. "Every space is interpreted differently by the different actors in it. The interpretation of space (and place) is based on subjective readings" (Davis 2001, 129-30). Indeed, it is the case that a single landscape cannot be the same for any two individuals because each has had a different interaction with the landscape, thus their knowledge of the landscape differs.

However, while the subjectivity of reading is complete, meaning does not fracture totally. To those commonly socialized by family, culture, and history, there are certain similarities in readings—there are congruencies and concordances. These unspoken institutionalized understandings, when entered into discourse, are then shaped, refined, and distilled by the inevitable power relations that are party to all human interaction. In this way, meaning in the landscape is just another example of what Foucault characterizes as power/knowledge. We do not mean to suggest that any landscape has universal meaning, or even that two individuals will be totally in agreement on the meaning of landscape. Rather, we mean to suggest that there is at least a grain of commonality or concordance in the readings of a great many landscapes and that discourse has a role in the attempt to further homogenize meaning. It is this concordance of meaning that tourism presupposes and relies upon (Lutz and Collins 1993).

Further complexity is encountered when we fully explore the meaning of landscapes for tourism theory since the interplay of concordance and geography results in meaning being geographically contextual. Davis (2001, 127) notes "histories, cultures, power relations, aesthetics and economics all combine at a place to create a context." It is simultaneously nature, habitat, artifact, system, problem, wealth, ideology, history, place, and aesthetic (Meinig 1979). Tourist objects and places (that is to say landscapes) have local, endemic, insider meanings and broader, pandemic, outsider meanings (Lowenthal and Prince 1972). For example, a pastoral landscape may inspire a romantic sentimentality in virtually any Western tourist, but knowledge of whose homes, farms and cattle are being gazed upon is restricted to the local inhabitants. Similarly, the birthplace of a local hero may mean nothing to outsiders. However, while insider meaning is place-bound, this does not mean that landscape has no meaning to those without local knowledge.

Zaring (1977), in her analysis of tourism in the Welsh mountains, reminds us that, in the absence of endemic meaning, we fall back onto that more general meaning of objects and places in our experience. She identifies two ways in which Western notions of aesthetics, based on late eighteenth century Romantic thought, interact with objects and places gazed upon—the Romantic sublime for "pure nature" and the Romantic sentimental for the historical and cultural. Resource-based tourism relies on Western notions of "pure" nature and its connection to the Romantic sublime wherein nature is represented as a spiritual domain for curing the ills of civilization (Hull and Revell 1989; Hyndman 2000). Conversely, a historic landscape hardly qualifies as "pure" nature. Thus, beyond a vague nostalgic sentimentality and yearning for simpler times, understanding of historic landscapes is necessarily limited to those who have "learned" the site (Pinder 2000; Lorzing 2001).

All this is to suggest that, far from tourism being simply a one-way process as suggested by the phrase "the tourist gaze," something far more nuanced happens in tourism. We argue that the locus of study for tourism is and should be the landscape. Tourism by definition takes place in a "tourism landscape." This tourism landscape, as Minca and Oakes (2006) suggest in their re-theorization of tourism using notions of place stemming from the humanistic tradition in geography, is more or less a reflection of the place being toured.[1] In this way the tourism landscape is the end result of a process of social construction that has played out over a number of decades and perhaps centuries and millennia. The tourism landscape may or may not be highly contested and it may or may not have been willfully constructed by a state that wishes to foreground certain attributes and background others. The act of touring is thus quite complex and revolves around deciphering the identity of a place and its inhabitants from that place's landscape, using all the tools available to the modern tourist (previous experiences, the internet, pocket histories, guidebooks, tour guides, and so on). While for some, tourism is still about getting away from it all or collecting "trophies" in the form of souvenirs from exotic places, even then tourism entails the deciphering of identity from clues in the landscape.

---

1   See also the edited volume by Ringer (1998) which offers case studies of tourist landscapes with a focus on the way in which local landscapes are changed by being reconceptualized as "destinations."

The Monograph

The chapters that follow examine the interconnections between identity and landscape, on the one hand, and landscape and tourism on the other. Chapter 2 provides a necessary introduction to the central themes of landscape studies. Greer, Donnelly, and Rickly point out that different observers seek to explain or interpret the landscape being studied from either the viewpoint of an external observer or that of a participant in the landscape itself. Meanings reflected in landscape studies fall generally into scientific and symbolic types, correlating generally with the explanatory and interpretive goals of study. Tourism activities are shown to build on various combinations of these complex parameters.

The next two chapters primarily focus on the identity-landscape connection and examine the manner in which historical struggles over identity have created landscapes that subsequently became foci for tourism. In Chapter 3, Huff argues that landscape is the reification of identity. Using the example of Strasbourg, France, Huff analyzes its landscape at the local, regional, national, and supranational levels, with particular attention to the manner in which place and identity interact, and what that interaction says about society. In Chapter 4, González-Vélez discusses culture's impact on one's ability to "read" a landscape with particular reference to the Copper Canyon region of Mexico. The region, its residents, and its external actors are studied with regard to the physical and socio-cultural changes that have taken place within the area as it has been opened up by the arrival of the *Chihuahua al Pacífico* rail line and the building of roads.

The next three chapters all examine the full identity-landscape-tourism nexus and focus on tourist landscapes that remain deeply problematic and, potentially, contested. In each case tourism has served to highlight certain aspects of history in the landscape and obscured other more problematic aspects. In Chapter 5, Soper studies the natural and cultural appeal of Mauritius to tourists, keeping in mind that the cultural dynamic of the island is changing. Giving a brief history of the island, she establishes Mauritius as a land of immigrants who have thus far used their beaches as the main pull for visitors. Mauritians' relatively new institution of "cultural heritage tourism" is discussed in detail, with special attention given to one site deemed crucial to the island's history and another that has been marginalized. In Chapter 6, Wolfel examines post-World War II culture in western Germany, and juxtaposes popular modernist and historical viewpoints about how a painful Nazi past should be treated. Highlighting the fact that many buildings needed to be rebuilt after the war, he explores the ways in which National Socialism was largely ignored during the re-culturation of Munich. Wolfel deals primarily with the city's architectural decisions and how the presence (or lack thereof) of monuments and other war remembrances might affect tourists' interpretations of German culture in Munich. In Chapter 7, Metro-Roland studies the manipulation of space and place in narrating history. She examines two sites dedicated to the remembrance of communism in Hungary, the House of Terror and the Statue Park. Locality within the city, the bounding of the narratives within each site, and the physical manifestations of historical reminiscence are explored in order to highlight their roles in shaping visitors' understanding of the past.

The final section more closely examines the landscape-tourism linkage from the differing viewpoints of insiders – participants in the landscape itself and outsiders – external viewers of the landscape. In Chapter 8, Timms uses the concept of parallax to examine the natural, social, and cultural landscapes of a Honduran park. Focusing on displaced residents from Celaque National Park, international NGOs given charge of the park's management, and tourists who visit the area, the author examines how differently insiders and outsiders can view the same natural landscape. Additionally, Timms touches on the broader issues related to the study of national parks by examining historical movements in the landscape tradition. In Chapter 9, Knudsen focuses on the history, culture, and natural environment of Denmark's Thy region, and its landscape's meaning for insiders and outsiders. He argues that outsiders cannot fully grasp Thy's many levels of significance, so they focus more on its natural/aesthetic aspects. Conversely, he explains how Thy's historical value renders it extremely important in Danish culture. In Chapter 10, Yespembetova, Rickly, and Braverman contrast the insider and outsider meanings of the Tamgaly rock paintings in Kazakhstan using four viewpoints drawn from landscape studies. Tamgaly is discussed as it relates to place, history, national symbols, and nature. Viewing the rock paintings as metaphors for Kazakh history, they explain how the petroglyphs are windows into Kazakh culture.

The final chapter summarizes the major aspects of the landscape approach to tourism. The notions that landscape is the reification of identity and that tourism is the quest to understand identity from the landscape of a place are re-examined. We close by contending that the landscape, identity, and tourism nexus is and should be the starting point for any theory of tourism.

.

Chapter 2

Landscape Perspective for Tourism Studies

Charles Greer, Shanon Donnelly, and Jillian M. Rickly

Introduction

Landscape research in the English language geographic literature has focused on the cultural landscape for much of the last century, leaving treatment of the non-human or "physical" landscape to other specialties in the discipline. The reasons for this include trends in academic research generally—the universal narrowing of research topics within the increasingly distinct realms of physical science and social science—as well as the intra-disciplinary dynamics of geography described by authors such as Butzer (1989) and Turner (2002).

Our purpose in this chapter is to provide context for the studies of tourism in subsequent chapters, which relate directly to the mainstream tradition of cultural landscape study in which cultural processes have garnered considerably more attention than physical systems of the environment or the composite physical-cultural landscape. We do not undertake the kind of comprehensive and more detailed survey of landscape research offered by Rowntree (1996) and Jones (1991).

Human Agency and the Cultural Landscape

Landscape study received major impetus from the work of Carl Sauer and the geography program which developed around him at the University of California at Berkeley. This approach was well established by the mid–twentieth century, and while it remained strongest as a research focus in universities of the western United States, it spread to other major research programs as well, evolving in subsequent decades through two major themes for elucidating the cultural landscape: first, development of the importance of human agency in shaping landscape features, and second, examination of the meanings for human culture systems embodied in built landscape features.

That an emphasis on culture systems, and especially that "agency" and "meaning" would eventually emerge as primary research themes, was not obvious when Sauer published his seminal article, "The Morphology of Landscape," in 1925. Sauer's formulation resulted from a combination of his grounding in European, especially German, geographic thought, and his own experience with field based studies through his early career. In a fundamental challenge to environmental determinism, the prevailing school of thought regarding the central geographic question of the nature of the human–environment relationship, "The Morphology of Landscape" asserted that complex interactions between physical and cultural influences combined to shape the geographic region. These relationships are clearly laid out in the oft quoted passage from Sauer stating, "The cultural landscape is fashioned from a natural landscape by a culture group. Culture is the agent, the natural area is the medium, the cultural landscape the result" (1925, 46). This conceptualization encompassed three elements with the potential to shape future directions in landscape study as the core of geographic research: "natural" elements, culture, and the resulting distinct area or region created by culture working on the natural landscape.

Sauer's 1925 scheme argues that geographers should seek to understand landscape as composed of areas or regions in the same way that the historian looks for eras in the process of studying the past. Areas are made up of important landscape features, just as eras are composed of prominent years, events, and personages. If we apply the term "cultural-chorological" approach to this viewpoint, it is precisely the one that was weakened and largely overpowered by the ascendance of "spatial-chorological" research after the mid–twentieth century (Turner 2002), so that Sauer's chorological component of landscape gradually became less emphasized and eventually all but lost. The same is not true, however, for the natural and cultural elements of landscape. These each have provided a basis on which landscape-related research themes have continued to evolve to the present time. But they have not been given equal attention, with the preference given to cultural elements exceeding that given natural elements to the point that the terms "landscape" and "cultural landscape" have become essentially synonymous in the usage of recent decades. In spite of being eclipsed in this way, however, non-human processes, which certainly remain important in shaping landscape features, are not entirely excluded from studies of landscape formation.

Ecology and Landscape

Darwinian-inspired ecological thinking emerged in the latter half of the nineteenth century, and in a short period of time influenced how scientists understood the relationship between humans and their environment. Ernst Haeckel (1899) and other early ecologists launched attempts to bridge the man–nature dualism that had widened with the burgeoning of scientific disciplines, and due to its interest in both human social systems and physical processes, geography was among the first of the sciences attracted to concepts from the emerging field of ecology. Some early geographers recognized that the non-human biological world and human socio-economic systems shared similar patterns of functional organization. Otto Schluter and Albrecht Penck, for example, "identified geography as the examination of the 'visible' landscape as a nature-society relationship" (Turner 2002, 57), but such an approach stayed well outside the mainstream of landscape research approaches in geography which was shaped by stronger tendencies within the discipline.

Most fundamental of these was the tradition of research on physical systems of the environment, led by prominent figures of the determinist approach including

William Morris Davis (1909) in geomorphology and Ellen Churchill Semple (1911) in climatology. The entrenched and separate foci given to physical systems and human systems ran counter to the holistic impulse of the nascent interest in ecology and were reinforced by the specialization and reductionism which shaped scientific pursuits throughout the twentieth century.

Also important to the failure of the ecological perspective at being accepted as a tool for landscape studies was Sauer's position that an ecological understanding could not satisfactorily explain human social systems (Zimmerer 1996), in spite of the important contention that his "morphological method" did apply equally to the natural and the cultural landscapes.

Calls such as those by Morgan and Moss (1965) and Stoddart (1965) for recognition of ecosystem principles as integral to the operation of geographic systems served more to save such a perspective from oblivion that to shape mainstream research trends.

Zimmerer (1996) finds five identifiable approaches which reflect the use of ecological perspectives within human geography. Generalizing from his argument, the important distinction between these approaches is whether or not relationships among humans can be described in the same terms as can relationships in the rest of nature. Zimmerer argues that three approaches, those identified as human ecology, systems ecology, and adaptive dynamics ecology, share the perspective that humans and nature are a single unified system which can be described by ecological and general systems concepts. These stand in contrast to the approaches of cultural–historical ecology and political ecology which apply ecological thinking to the non-human components of the system while applying social theory to the human components. Looking back at the roots of early ecology, Zimmerer suggests that the difference may largely boil down to whether human social organization can be understood in terms of cooperation as an adaptive strategy or whether it is some sort of moral or ethical capability unique to the human species.

Application of ecological perspectives with implications for landscape studies in geography is led by cultural ecologists including Nietschmann (1973) and Butzer (1976). Other discussions and applications have come from Chappell (1975), Demeritt (1994) and Birks et al. (1988), while the variety of approaches is well surveyed by Mathewson's essays (1998; 2000). The voluminous work of Pierre Dansereau (e.g. 1975, 1983) deserves much more attention than it receives in this regard, for the perspective it brings to a sophisticated ecosystem approach to landscape which includes the human role.

The developing field of landscape ecology also has been particularly relevant to landscape studies. The contents of landscape ecology as a field, however, are quite different in the European and North American contexts. The reasons for these differences are debated, but in a review of the intellectual roots of landscape ecology, Turner et al. (2001) suggest that the older tradition of landscape ecology in Europe focuses on planning and management of the landscape because the land of that region has been much more intensively managed for a much longer time. In contrast, landscape ecology in North America is a much younger field and has taken a more biological perspective, more often focusing on the pattern-process link between habitat and non-human organisms. The historical development of landscape ecology in Europe (as reviewed by Naveh and Lieberman 1994, 21) again highlights the rich meaning of landscape as "the total spatial and visual entity of human living space".

Another recent research area with influence on landscape studies is political ecology, combining ecological perspectives and critical social theory to focus on issues of environmental justice such as the location of toxic waste dumps (Pellow and Brulle 2005), food security (Unruh, Heynen, and Hossler 2003), differential access to environmental amenities (Heynen, Perkins, and Roy 2006), and equity in environmental policy (Desfor and Keil 2004).

The Cultural Landscape

Although Sauer's morphologic method was intended for application equally to "natural" and "cultural" features, and therefore Agnew, Livingstone, and Rodgers (1996, 234) are justified in their statement that "modern geography began as an experiment in holding nature and culture in the same analytical framework," the method was not equally applied, and that experiment did not hold together. The construct put forward in "The Morphology of Landscape" was a casualty of forces emerging through subsequent decades in geography. First, an intellectual environment developed hostile to landscape as a legitimate research theme (see Butzer 1989 and Turner 2002). Secondly, Sauer himself moved in a direction that resulted in giving culture the momentum of a pendulum swung against the determinists' previous emphasis on natural forces as the dominant force in the human–environment relationship. Sauer's extended influence on the discipline came from an evolution of concepts through his career, indicated by signpost publications which followed "The Morphology of Landscape" (1925), with "Forward to Historical Geography" (1941) and "The Agency of Man on Earth" (1955). Emphasis on the temporal or historical element of landscape evolution, applied increasingly in a variety of New World field work settings, resulted in Sauer's "agricultural origins and dispersals" theme which had at least as much impact outside geography as within. The "human agency" theme coalesced and emerged from this work, giving Sauer and his students' prominence in the multi-disciplinary 1955 conference on "Man's Role in Changing the Face of the Earth," a platform for the "human impacts" point of view basic to the broad and multi-disciplinary environmental movement which ensued.

The effect of this evolution was to shape a good deal of research in geography more by perspective and world view than by explicit methodology. While some landscape research represented quite conscious implementation of the morphologic method (Broek 1932; Aschmann 1959; Nelson 1959; Raup 1959), more common was wider-ranging work of later decades which retained the perspective and the spirit of the human agency approach, but which adopted more focused or specific questions by individual authors, such as house types (Kniffen 1965), culture regions (Jordan 1970), architectural features (Groth and Bressie 1997), rural landscape features (Hart 1998) and the sites of tragic or violent events (Foote 2003). Equally important was the reinforcement which the human agency perspective gave to landscape studies outside the discipline of geography illustrated by the work of J.B. Jackson (1970) and environmental historians such as Conzen (1990), Stilgoe (1982), and Cronon (2003).

Culture was thus established as the province in which landscape research was pursued in geography. While other specialties in the discipline took economics or mathematics as cognate fields from which to develop tools for shaping their research, landscape studies found affinity with anthropology and folklore. In such borrowing, however, it tended to be single factor elements which were employed, rather than conceptually complex relationships within the cognate field. In borrowing from anthropology, for example, it was not the interrelationships among technological, organizational, and cognitive patterns of society which were imported for analysis of their impacts on landscape. Instead, it was house types, field patterns, or other individual culture traits which were used to study diffusion, sequent occupance, or other topics of interest to geography. How selected traits found expression on the land was more important than how the complex societal system works, or even less, how that system interacted with physical systems of the environment.

The effect of this evolution away from the tenets of the construct presented in "The Morphology of Landscape"—time acting on the complex of both natural and cultural features produces regions of the composite landscape—was to remove modern landscape study from the longstanding tradition in geography focused on understanding how humans and their environment interact to shape the different places on the surface of the earth. Cut loose from this human–environment anchor point, landscape studies have been free for several decades to follow the paradigms of culture studies.

Landscape Perception and Behavior

The methods of environmental perception came into geographic studies as part of the environmental movement of the 1960s and 1970s, impacting both the social science and humanist traditions in the discipline. For landscape study, which was firmly established as a theme in cultural geography, this opened the way for examination of connections between landscape's dual identities as both a recognizable portion of the earth's surface and a subject of representational art. The evolution of these two identities side-by-side in the Germanic culture region of northern Europe has been well presented by Olwig (1996).

David Lowenthal (1961) led the way with his examination of John K. Wright's Geosophy, or imagination in geographic knowledge, and YiFu Tuan (1971, 2003) developed a career bringing cross-cultural perspectives to this point of view. The evolution of humanist approaches to the present time is well illustrated with Olwig's (2002) examination of ideology and landscape.

Lowenthal and Prince (1965) took the next step in linking collective cultural values with built features of the landscape as a mutually reinforcing feedback loop, in a way not attempted by the human agency approach to landscape, but limited to use of selected features of the landscape rather than the composite of all features as pursued in traditional approaches up to and through the statement of Sauer's morphologic method. The emphasis on cognitive patterns in this approach also leaves unexamined with equal scrutiny how these patterns connect with institutional and technological elements of a culture system to interact with physical systems of the environment in shaping landscapes (Tuan 1968).

It can be puzzling that natural hazards research, a practical, planning-oriented endeavor, is discussed in connection with the humanist tradition (Rowntree 1996) until the point of view on which it is based becomes clear. The development of this work by Gilbert White and his students (White 1945; Kates 1962; Saarinen 1966) employed the perspective of environmental perception in societal adjustment to flood and other processes which impact human settlement. Based on consideration of interactions between society and the physical environment, and not always from a simple, single factor point of view, the work would seem to necessitate an objective, composite landscape point of view. But the degree to which focus on culture had established an anthropocentric basis for research is reflected in the heading "natural hazards" itself. Environmental processes driven by energy systems operating at the earth's surface for the duration of the planet's existence are perceived as hazards from inside human society looking out, increasingly so with greater encroachment by human development. Such focus on perception implies a viewpoint on human/ habitat relations in which environmental processes are secondary to processes of human society.

This and related viewpoints created a context for new directions in landscape study through the next several decades. Reliance on subjective understandings and culture system operation suggested by Donald Meinig's "The Beholding Eye: Ten Versions of the Same Scene" (1979) and Pierce Lewis's "Axioms for Reading the Landscape" (1979), moved perception to the mainstream of landscape geography via application to the burgeoning examination of vernacular landscapes. Concurrently, the stage was set for "nature" to be understood as concepts differently constructed by different culture groups (see Simmons 1993), that is, as phenomena of cognitive patterns in human culture systems rather than as forces operating on surface features of the earth since before any life forms appeared, including hominids and the culture systems they have evolved.

Meaning and the Cultural Landscape

The shift from approaching cultural landscape as the product of human agency to its being the repository of meaning for human societies, suggested by Meinig and Lewis, led to an increased focus by researchers on symbolism and iconography (Penning-Rowsell and Lowenthal 1986). "All landscapes are symbolic," argues Cosgrove (1989, 125), as they are deposits of collective cultural and symbolic meaning (Daniels and Cosgrove 1988; Cosgrove 1989). Rowntree and Conkey (1980, 474) suggest that cultural geography in the past had seen cultural landscapes "as only a reflection of social process ... [h]owever, intangibles, such as social identity are, in fact, *realized* by the landscape." That is, the intangibles are made real by landscape, illustrating the feedback between built features and collective patterns of cognition.

This linking of perception, symbolic landscapes, and social theory was pursued by Cosgrove (1984) who drew on the aesthetic tradition of landscape painting. He argued that a certain way of seeing the land evolved from Renaissance Europe, whereby stylized images were used to exert the power of class and social hierarchy; and he observed that this way of seeing, developed in urban contexts, was then exported to rural areas under the control of the city. Similarly, Osborne (1998, 452) argues that power "is asserted by the exclusion of, or transformation of, the commemorative practices of others." In a fuller development of the theme, Cosgrove (1986) draws more detailed links between modes of production, models of capitalist transition, and tensions between culture and landscape. Such affinities for cultural studies and cultural theory naturally resulted in the variety of perspectives which evolved in the late twentieth century being applied in geographic landscape studies. Authors such as Duncan and Duncan (1988) built on the idea that landscapes can be read as texts that give ideologies a concrete form and as such, can be analyzed with textual analysis methodologies.

If landscapes can be read as ideologically laden texts in concrete form, then it is important to consider the notion that landscapes are made for particular purposes. Lefebvre (1991, 31) notes that "that every society ... produces a space, its own space." Among the research themes launched by this perspective are several with relevance for landscapes of tourism: landscape aesthetics, landscapes of power, and especially insider/outsider meanings of a place or a landscape. The first of these, aesthetics, as the juncture of perception, values embodied in ideology, and purposeful creation of space, provides an example of the cultural landscape/tourism nexus worthy of some elaboration.

Zaring (1977) posits that examination of the production of landscape in Europe yields three iconic landscapes as symbols for society—the beautiful, the picturesque, and the sublime. These three are the result of an aesthetic developed in the eighteenth century. From the time of the Enlightenment, wherein reason triumphed over belief, until the rise of Romanticism in the eighteenth century, the prevailing aesthetic was that of the beautiful. The beautiful, which was typified by the rigid use of perspective, continuous line and symmetry, was the logical outcome of divine reason and absolute monarchy. Romanticism changed that by adding two additional aesthetics—the sublime and the picturesque. The sublime champions the power of nature (or God) over all humanity. The picturesque is an aesthetic moment which is grounded in the landed aristocracy and mercantile class of the British Empire (Knudsen and Greer 2007).

The beautiful values the "neat, clear, harmonious, and self-contained forms" of the classical era and stresses the role of human agency in their creation (Schönle 2000, 598). The sublime values the "uncontrolled and chaotic growth of nature" and stresses the "contemplation of an object of such greatness, darkness and violence" that it overwhelms our comprehension (Schönle 2000, 598; Knudsen and Greer 2007).

The picturesque rejects the beautiful as too perfect, too artificial, and too divorced from nature, but also rejects the sublime as too "wild, threatening, and uncontrollable" (Schönle 2000, 598). It is a mediation of the beautiful and sublime in that it is the contemplative moment experienced when one is both attracted, as to the beautiful, and awed, as by the sublime (Schönle 2000, 597; Knudsen and Greer 2007).

As an aesthetic the picturesque has its origins in the theater—it is a method of seeing predicated on both framing and distancing (Marshall 2002). Framing turns landscape into *a scene*. In the same way that viewers in the theater are spectators and a play is clearly meant to be looked at, the viewer of the picturesque treats the landscape as a sequence of views. The need to frame is exemplified in the widespread use of the Claude glass (named after French artist Claude Lorraine) by eighteenth-century tourists when touring, which required that the viewer have her back to the actual scene being viewed, and by the importance placed on viewing through windows and from prospect points (Marshall 2002). And while the composition of the picturesque is facilitated by framing devices, the picturesque is, first and foremost, a way of internally imposing on nature the proscenium of a theater stage or the frame of a painting (Marshall 2002). Through the picturesque, a "way of *looking* … [becomes] a way of *knowing*" (Barrell 1972 quoted in Marshall 2002, 415, italics added; also Schulenburg 2003). The picturesque "depends on providing views and scenes to a spectator from some privileged vantage point" (Townsend 1997, 370; Knudsen and Greer 2007).

The importance of recognizing such aesthetics as collectively constructed images goes beyond their use for galvanizing social cohesion and for implementation in iconic landscapes from gardens, formal and informal, to national parks. Aesthetics represent an untapped potential for illustrating the abstract societal ecosystem–landscape nexus put forth in Dansereau's (1975) *Inscape and Landscape*. That monograph posits an essential role for collective perception of environment as an initial step in a social group as ecosystem element first conceiving then creating a landscape which it inhabits. The ultimate integration of humanist, social scientist, and physical scientist contributions to understanding of landscape will depend on such linkages being recognized and jointly pursued.

The perspective of landscape as power derives from the point made by Duncan and Duncan (2001), among others, that landscapes mean different things to different people and are therefore heterotopic (see Foucault 1986). This establishes landscape as an entity whose meaning is derived from discourse and the inevitable contestation that surrounds it (Foucault 1980). Schein (1997, 663) suggests that "when particular landscape readings prevail over others, the places on a landscape become discourse materialized." Landscape does not "merely signify or symbolize power relations; it is an instrument of cultural power. [It] exerts a subtle power over people, eliciting a broad range of emotions and meanings that may be difficult to specify" (Mitchell 2002, vii, 1). It is in the discourse over the meaning of landscapes (among other places) that the power relations in a society reveal themselves, with power in this case being the ability to shape places, events, and perceptions. These power relations in the landscape frequently can be studied by comparing insider and outsider meanings of a place or landscape.

As the terms suggest, an insider is someone who occupies a landscape, whereas the outsider does not. Cosgrove (1984, 19) states, "[f]or the insider there is no clear separation of self from the scene" because place is embodied with a personal and social meaning. Because the insider and the outsider experience landscape very differently, this difference often is a salient aspect of any discourse over the meaning of a landscape. Traditionally, geographical inquiries into landscape have for the most part used the perspective of the outsider for understanding the meaning of landscape form and features. However, understanding meaning and meaning contestation, and therefore power and landscape, requires understanding the insider perspective, which is far more laden with meaning than is outwardly visible (Cosgrove 1984).

Finally, it is important to note that the insider/outsider divide can be broken down into race, gender, and class categories. The category of race is the result of generalization as humans try to group people by similar traits, such as skin color or ethnicity (Anderson 1987). This is a category of meaning contestation that conjures deep, historically rooted conflicts. Acts as seemingly simple as changing the name of a landscape commemoration feature can elicit a contest in which centuries of conflict become symbolized by the one moment of feature renaming. Several authors including Leib (2002), Hoelsher (2003), and Dwyer (2004) have addressed the role of race in landscape studies.

Most literature dealing with gender, whether it is within the realm of landscape studies or not, distinguishes between sex and gender. In this context, sex is commonly defined as "the biological differentiation of male from female," whereas gender is associated with "the cross-cultural and historically varying expressions of masculinity and femininity" (Monk 1992, 123). Gender and power contestation can easily be seen in most landscapes, as addressed by Yeoh and Huang (1998), McDowell (1983), and Miller (1983).

Class, the social structure of a community, is based on age, income, or similar elements that group people together. Contestation by class also deals with topics such as conservation versus development (see Zukin 1991; Mills 1988; Crump 1999; Patano 2005; Sletto 2005) and contestation within colonial landscapes (see Leitner and Knag 1999; Whelan 2002; van Eeden 2004). Authors such as Duncan (1993), Schein (1997), Duncan and Duncan (2001), Mitchell (2001), and Orser (2006) focus their studies on class and power relations in the landscape.

Conclusion

The studies which make up this book rely on themes of the cultural landscape as starting points for linking the tourist place or activity to various expressions of meaning. While these chapters explore recent and contemporary themes, we hope to bring light to the fact that the range of themes through which cultural landscape studies has evolved can be seen as links to tourism. Obvious pairings include: perception and the tourist gaze, symbolic landscapes and heritage tourism, and ecological relations with eco-tourism.

The notion of the symbolic landscape may be the most applicable of the landscape perspectives to tourism studies, especially within the increasingly popular "tourist landscape" literature of which this book is a part (see also Aitchison, Macleod, and Shaw 2001). Most aspects of the symbolic landscape easily translate to tourism studies. It is obvious to see how notions of insider (receiving society) and outsider (tourist) consciousness, and variations in their interpretations, play out in the tourist landscape. However, the power relations of identity (and the categories of race, gender, and class) are also visible in the tourism landscape. Meaning contestation occurs in every landscape, tourism landscapes are no exception.

Chapter 3

Identity and Landscape:
The Reification of Place in
Strasbourg, France

Sean Huff

Introduction

Tourists cannot be tourists without landscapes to tour. While it is the role of tourists to tease out the identity of places from their landscapes, the role of this chapter is to investigate more closely the relationship between landscapes and identity using Strasbourg as an example.

When visiting Strasbourg, the tourist confronts an urban landscape that is simultaneously regional, national, and supranational. Strasbourg developed slowly throughout the past 3,000 years, during which time it had been Celtic, Roman, German, French, and currently it is a unique mix of Alsatian, French and European (Dreyfus 1979). Thus the tourist can expect to see evidence of Strasbourg as Alsatian city, as contested national territory, and as supranational metropole in the urban landscape of Strasbourg—one can expect to see the reification of the multifaceted identity of the city in the city's landscape.

Alsace, the region in which Strasbourg lies, is the smallest region of France. It has historically been at the crossroads of Europe and the line of demarcation between French and Germanic culture. Linguistically and culturally distinct from the surrounding states of France, Germany, and Switzerland, Alsace remains a unique cultural region within an increasingly international Europe.

Historically, the identity of Europeans was formulated around different configurations such as city-states and regions, but the importance of regions declined with the emergence of nation-states (Fowler 1952; Mackenney 1989; Sakellariou 1989). As nation-states emerged, Alsace became a frequent bone of contention between France and Germany.

More recently, the forces of globalization, namely supranationalism in the form of the European Union, have been felt throughout Europe, and it is becoming clear that regions, specifically European regions, are once again increasingly important both politically and culturally (Applegate 1999). They are engaged economically, both nationally and internationally, independent of their respective states (Harvie 1994). Alsace, with Strasbourg as its major city of 252,000 people, is on the cusp of this cultural and political change in Europe. It is the home of the European Parliament, the Council of Europe and numerous multinational organizations. This geographical heart of Europe, traditionally a dividing line between warring nations, is becoming a major center of a supranational network.

In what follows, the relationship between the identity of a place and the landscape of a place is examined. This is followed by a discussion of the manifestations of Strasbourg as regional center, contested national territory and supranational city. The chapter closes with a summary and conclusions that return to thoughts about how tourism involves the deciphering of the identity of places from their landscapes.

Landscape, Place, Identity: A Synthesis

Because this analysis is focused on landscape, the idea must be fully understood. Landscape is not merely the lay of the land or a type of painting but a way in which to conceive of place. The term "landscape" is contested in geography because, according to Cosgrove (1984, 13), it is "an imprecise and ambiguous concept." Clearly landscape is not only a place, or a collection of things in a prescribed area, but also a way in which to conceive of place—it is the canvas on which the social constructions of place are reified in order to create and maintain the identity of a place. Landscape is not static, but active and ever changing. Each generation of society manipulates their landscape to represent their ideas and identity which results in a palimpsest of continually overlaid landscapes. As a result, landscapes must not be "wrenched out of [their] context of time and space" (Cosgrove 1989, 127) but rather read as pages of a cultural text, each distinctive from the next.

Landscape is the background of our collective existence underscoring our identity (Jackson 1984). Within the landscape, identity is manifested in physical symbols. Architectural style, urban design, statuary and public spaces are all infused with the identity of a people. These symbols "serve the purpose of reproducing cultural norms and establishing the values of dominant groups across all of a society" (Cosgrove 1989, 125). In other words, societies construct symbols to represent how they understand themselves (Harvey 1979).

Although places are social constructions and home, nation, and region are all subjective ideas existing in social discourse, places, with their unique and distinctive qualities, help to create an identity for society. Places are made real through the visual. Abstract notions, such as that of nationhood, only become real when the iconography of nationhood is created, given meaning and peppered throughout the landscape. Buildings, memorials, statues, and painted representations of the national landscape are more than mere constructions of a society, but rather they are symbols of social identity, created and maintained to perpetuate the idea of place. Place is shown to be real by real representations of it. It is this idea that is one of the focal points of this research.

Identity, defined here, is a sense of self derived from lived experiences and subjective feelings embedded in social relations (Rose 1995). Identity is a complex idea, constructed on the basis of language, religion, civic values, religion, ethnicity, myths, historical memories, and place (Smith 1991; Beetham and Lord 1998; Graham 1998b; Dijkink 1999; Paasi 2001). These variables of identity formation are internalized from lived experiences in a community. Identity has been directly linked to territory or the conception of territory in the past, but that association is being threatened in the modern world (Harvey 1989; Hall 1995).

Graham (1998a, 2) sees evidence that "identity is not a discrete social construction that is territorially bounded; rather, identities ... overlap in complex ways and geographical scales." Though multiple identities exist at once, they cannot be considered hierarchical. They are only hierarchical when considered contextually. According to Graham, national identity is not more important than local identity except when placed in a context such as war or diplomacy. Identity can therefore exist in many forms and for many reasons. Political, linguistic and religious identities may all exist within one community, overlapping one another in a complex patchwork. The transformation of modern identity must be considered in this context to allow for comparison and recognition of change.

Linking landscape and place leads to a greater awareness of the concept of landscape as a manifestation of ideology, culture, and identity. Analysis of landscapes as representations of local, regional, national, and supranational identity forces one to think about our constructions of identity around historically contingent places such as regions and nations. Landscapes cannot be segmented and calculated, but their construction, orientation, preservation, and deconstruction aid in our understanding of the relationship between humans and place.

It has been said that the relationship between identity and territory is tenuous in the modern age as places are commodified and sold in a time-space compressed world. Nevertheless, identity is closely related to the territory of a society, as the culture and identity of a society is manifested and perpetuated in its landscape. Landscapes are the concrete manifestations of conceptions of place and as such, they represent the cultural values of a group or people.

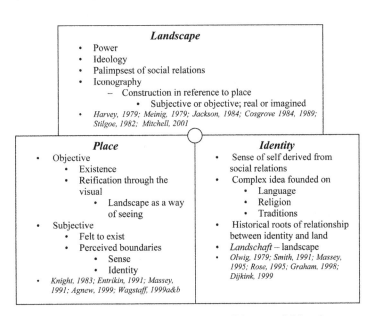

Figure 3.1 Theoretical Synthesis of Landscape, Place and Identity

The analysis of Strasbourg, France, to follow, is concerned primarily with this intersection of landscape, place and identity (see Figure 3.1). Landscape, place, and identity are each complex ideas. For example, landscapes are more than monuments and urban design, they are also reflections of our treatment of the environment as well as an example of how religion influences our perception of land and resources. Places are more than states, regions and cities. Identity is infinitely complex and is expressed in a multitude of ways. Therefore, it is important to highlight the fact that this study is focused on the intersection of these complex ideas in an attempt to better understand that interaction, rather than cover the entire array of ideas within each individual variable.

A Methodological Note

Strasbourg, with its long history of evolution and change, has seen a shift in dominant identity from regional and national to a more local and supranational one. Though clearly concepts as fluid as identity and place are not mutually exclusive, a subtle transition appears to be occurring. This study examines the transformation of landscape iconography from German and French nationalism to Europeanism and covers the period from 1870–2002. The methodology used here traces this transition by recognizing two main historical periods of identity of construction and preservation: 1870–1970 as primarily regional and national and 1970–2002 as primarily local and supranational.

Of course the landscape in its entirety cannot be studied in the short space allotted here. As a result, this analysis will utilize three classic features of landscape which feature in the literature to present the evidence of shifting identity structures in Strasbourg, France: urban design (see Cosgrove 1989; Jellicoe and Jellicoe 1975; Stilgoe 1982), architectural style (see Meinig 1979; Cosgrove 1989), and statuary (see Harvey 1979; Atkinson and Cosgrove 1998; Osborne 2001).

The Regional and National Period: 1870–1970

The Strasbourg landscape embodies strong regional and national markers dating from the period 1870–1970. Occupied by the Germans in 1871, Alsace became a French territory again after the First World War. It was reoccupied by the Germans during the Second World War and reunited with France at the end of the war. The urban design, architectural styles, and statuary all bear witness to Alsace as a contested national territory.

This section considers two aspects of the regional and national period of Strasbourg's history. The first is the Strasbourg extension and university, examples of urban design from the era of German control after the Franco–Prussian war. The second is statuary in Strasbourg that commemorates military heroes from this portion of Strasbourg's history.

German Nationalism, Urban Design and the Strasbourg Extension and University

An analysis of urban design provides vital clues in linking the design of the built environment to social constructions of place, as a means to uncover the shifting orientation of social identity in Strasbourg, France. Major features of urban design in the study area are analogous to the identity of Strasbourg society's orientation toward regional and national constructions of place. Two primary examples are the Strasbourg extension and the construction of the national university, each representing a turning point in the representations of place construction in Strasbourg.

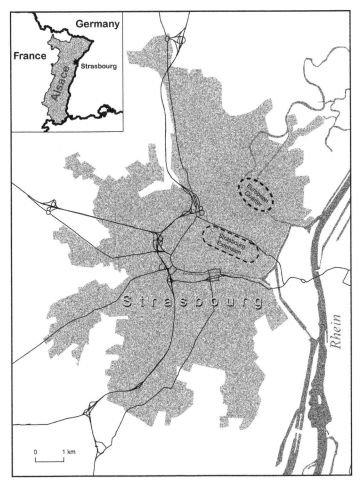

Map 3.1 Strasbourg, France (by Shanon Donnelly)

Following the occupation of Alsace and Lorraine in 1871 by Germany, a plan was put in place to create a Strasbourg extension north and east of the town center (Wilcken 2000). Strasbourg was to become the grand capital of the *Reichsland* region of the newly formed political union of Germany. The German political elite was reclaiming German territory lost in the signing of the Treaty of Westphalia in 1648, where up until at least 1910 a majority of the population still spoke a dialect of High German (Boswell 2000). German occupiers sought to incorporate ancient Alsatian constructions into their modern design rather than begin a process of complete landscape alteration.

The design began on the traditional roland of Strasbourg, the Notre Dame Cathedral in the heart of the ancient town center (see Figure 3.2). Stilgoe (1982) outlined the concept of roland as the religious and cultural center of a town, manifested in a physical object. The location of the Cathedral in central Strasbourg is the highest point on the central island and has served as an animistic religious ground, Roman temple, and Christian cathedral.

Figure 3.2 'Roland' of Strasbourg with the Cathedral

The centerpiece of German construction was labeled *Bismarkplatz*, honoring the great unifier of modern Germany. Lining this central *platz* were the centers of regional power in Strasbourg, designed to supplant French and Alsatian identity with a new German national identity for the Alsatian people. This "Imperial Forum" (Nohlen 1997, 41), housing the library, theatre and central authority, sat at one end of a grand avenue which terminated at the opposite end with the newly constructed university. This forum represented the needs of Berlin (Nohlen 1984, 162), but also the axis from which a "liaison" or connection to the old town of Strasbourg was formed. Clearly this aggrandizement by the German state sought to fill the new capital of the *Reichsland* with the iconography of statehood and nationalism while ensuring control over the landscape.

The most important factor in this landscape alteration was the consideration of Alsace as historic German territory, merely reclaimed by the newly powerful German state. Considering this, the amelioration of existing Alsatian constructions rather than the wholesale construction of a new city makes sense. The roland of ancient Strasbourg, the Notre Dame Cathedral, which was the center of the Strasbourg city-state during the Holy Roman Empire, was to serve as the roland of *Straßburg* through orientation and design schemes.

In the aftermath of the Franco–Prussian War Strasbourg was to become the home of a new jewel in the German higher education system with the organization and construction of the University of Strasbourg, which was situated at the opposite end of the grand avenue. It was thought that Alsatian identity, in its true German form, could be realized if leadership of the German state instituted educational and cultural policies aimed at its reconstitution. The construction of a new university in conjunction with the other grand features of the Strasbourg extension could facilitate the integration of Alsace into the German state, and assimilation into German culture.

The University of Strasbourg, linked with *Bismarkplatz* through the grand avenue, provided the liaison to the ancient town center of Strasbourg. It is a clear example of how urban design, the placement of important structures within the landscape, is not merely a method of arranging people and resources for efficient movement, but rather a powerful tool used in the reification of our subjective ideas of place in landscapes. The German leadership understood and utilized this fact of identity creation in their occupation of Alsace in the late nineteenth to early twentieth century. Both they and the French also understood the power of statues as symbols in the landscape.

Symbolism and Statuary, 1870–1970

Statuary and monuments have the role of mnemonic devices to perpetuate identity, embodying certain ideas, beliefs, traditions, or historical memories. Because they are so loaded with meaning and contentious they are difficult to study. One way to study this is to focus on the debates revolving around the completion or construction of a monument. An understanding of how the reification of place occurs through statuary can be garnered from looking at two famous statues from the center of Strasbourg, that of General Kléber and General LeClerc.

Perched atop a pedestal in the middle of the biggest *place*, or public square, in Strasbourg is a bronze representation of General Kléber (see Figure 3.3). A native

**Figure 3.3 Statue of General Kléber Located in Place Kléber,
          Central Strasbourg**

of Strasbourg, born in 1753 on the square which now bears his name, he served
Napoleon Bonaparte as a general from 1793–1800, when he was killed in Cairo. In
1840 the monument seen today in *Place Kléber* was completed by Philippe Grass
and erected to commemorate Kléber's life and his service under Napoleon.

   This representation of a local hero has come to represent a powerful ideal of
French nationalism in Strasbourg. This occurred primarily for two reasons. First,
Kléber served meritoriously in the French Army for Napoleon during the formative
years of the French Republic. Second, and most importantly here, the statue now

located in *Place Kléber* was removed during the Nazi occupation of Strasbourg from 1940–1944 and interred out of sight in the history museum. This simple act of altering the landscape of Strasbourg by the Nazis ensured the continued importance of Kléber as a symbol of French nationalism. He immediately became a symbol of freedom, of being French, and of liberation from Nazi domination during the Second World War.

Following World War II, and as a response to the Nazification which took place during the war, the statue of Kléber was placed once again in its proper location to represent the strong connection of the region to the French state. No longer did Alsatians wish to exist on the fringe between two warring states, they wanted to wholeheartedly join one. While one might say that France was merely an ideal, the ability to see and touch an icon of French national identity helped citizens of Strasbourg reconnect and reassert their connection to the French state.

Figure 3.4 Statue of General LeClerc in Place Broglie, Central Strasbourg

Another monument located not far from that of that of Kléber in Strasbourg is that of General LeClerc (see Figure 3.4). Located in *Place Broglie*, the statue is a memorial not only to LeClerc, but also to the lives of those soldiers lost in World War II during the battle to liberate Strasbourg and Alsace from the Nazi aggressors (see Figure 3.4). This example of statuary, more than most things in Strasbourg, represents the multiplicity of identities of the city. It commemorates General LeClerc, the end of World War II, and it honors those who gave their lives during the war, as well as highlights the symbolic reclamation of France from Germany and of Alsace by France. Yet the statue is also deeply ironic in that during World War II, numerous Alsatians fought for Nazi Germany against the Allied forces.

The Local and Supranational Period: 1970–2002

The late 1960s and early 1970s were the formative years of the burgeoning European Union. Strasbourg's extended history as the crossroads of Europe proved vital in the decision to place symbols of European unification, such as the European Parliament, within its landscape. Strasbourg began its climb from a frontier town of the French Republic, and indeed even the capital of Alsace, to a European capital city representative of the new union. Simultaneously, Strasbourg increasingly reasserted its identity as first and foremost an Alsatian city.

Construction of the New European Capital

In 1975, the *Palais de l'Europe* [the Palace of Europe] was completed in Strasbourg to house the Council of Europe. The Council building was the first major symbolic construction in Strasbourg of a newly reconstituted Europe after World War II. For Strasbourg, the Council building also represents the beginning of the attempt to construct a new, supranational European city. Strasbourg, with its varied history of control, allegiance, and orientation proved to be coming into its own. It increasingly came to symbolize neither France nor Germany, but a greater Europe. The Council of Europe, housed in Strasbourg, became the self-proclaimed first international parliamentary assembly in history, and its building is symbolic of Strasbourg's position as neither wholly French nor wholly German, but the potential supranational European future. Strasbourg represented the nexus of Franco-German contestation and was therefore a fitting location for symbols of reconciliation.

Following World War II, the idea of being "European" really did not exist in the minds of Europeans as perhaps it did in the minds of Americans. It has always existed in discourse as some vague notion of cultural togetherness, but quickly disintegrates with any specific comparative analysis. Do Wales and Catalan truly have much in common? Following the devastation of World War II, it became apparent that the creation of a European alliance was important to prevent future aggression and conflict. The Council of Europe was a first attempt towards making "the idea of Europe" a reality.

Figure 3.5 European Parliament Building (by Elisabeth Epstein)

As the European Union (EU) emerged, Strasbourg continued to garner new constructions including the European Parliament in 1993 and the European Court of Human Rights (see Figure 3.5) in 1995. During this period, the rejection of modernity, which signaled a rejection of the corruption and traditions of old Europe, was to inculcate a new generation of Europeans into the post-modern, internationalist age. The builders of futuristic, "high tech" European constructions such as the Parliament and the Court of Human Rights rejected the style of modernism, in which the right angle predominated producing tall, straight, box-like structures where form followed function (Harris and Lippman 1986). Instead they were to speak with curves, domes and form. "Architecture Studio," the architectural team responsible for the design of the new Parliamentary building, sought to construct a truly European building, symbolizing no one culture or tradition, but instead projecting an international idea.

The struggle for the placement of the European Parliament within the confines of Strasbourg reflects the ongoing struggle for the basic formation of the European Union itself. The fierce arguments regarding the placement of EU institutions have led to the discussion of the creation of a European quarter in Strasbourg surrounding the numerous constructions, analogous to Vatican City or Washington, D.C. Though no official exterritoriality exists for the European quarter as it does in Rome, construction has remained clustered in this area, with the possibility that a change in status could happen in the future. The construction of European buildings has led the populace of Strasbourg to compare themselves to New Yorkers and Genevans, inhabitants of cities housing many international organizations without being capitals in their own states.

The style of the current constructions is perhaps more difficult to analyze. They represent a vision of post-modernity peculiar to the new Europe. The European Union constructions in Strasbourg, designed to make manifest the new idea of Europe as an economic and political union, are overtly post-modern and acultural. They purposefully do not represent values or cultural roots of any one nation, people, state, or ideology, but merely the iconography of supranationalism and Europeaness. The European landscape in Strasbourg clearly separates itself from an iconographic past of fascism, communism, and war and atrocity. Post-modernity, in its lack of symbolism, is patently vague. One of the charges laid against the Union is its lack of European symbolism—no common language, culture, or tradition exists linking the divergent European countries together. Indeed, the only tangible symbol indicating that the EU is indeed real is the Euro, the currency of the Union.

Strasbourg is unique in this regard because of its urban landscape in which "Europe" has been reified through the construction of the European quarter. In Strasbourg "the idea of Europe" is realized because numerous symbols of it exist in the urban fabric of everyday life. The lack of common European iconography *within* the buildings is less important for the residents of Strasbourg because the overwhelming presence of the buildings and the 12,000 employees who inundate the town periodically prove Europe's existence. However, mention of the periodic influx of "Eurocrats" also raises the question as to how these "tourists" interpret the landscape of Strasbourg. Does the city represent "Europe" for those Eurocrats coming from Wales or Catalonia or does the city seem uniquely French or Alsatian? Most likely it is both.

However, in the villages of southern Alsace the idea of Europe gives way, as it does elsewhere in Europe, to nationalism. Here is where the lack of European unity, the lack of a real place called Europe in the consciousness of Europeans themselves, becomes blatantly clear. Lack of common culture, traditions, politics, economic principles and even agricultural practices set people apart. A lack of meaningful symbolism perpetuates the idea of difference among the people of Europe. This highlights the important idea of the contextuality of landscape. Within Strasbourg the European constructions represent an idea of wealth, prosperity, and internationalism, while outside of Strasbourg these European constructs represent a massive, out of touch bureaucracy. This feeling is often directed at Brussels and all other major features of the European Union bureaucracy within Europe today.

The profound interrelation of symbolism and identity creation cannot be easily glossed over. Historical approaches to identity creation in a multi-cultural environment show that symbols are profoundly important. The creation of the French nation involved a multifaceted Frenchification program including political organization, cultural homogeneity, perhaps the most effective being the organization and standardization of the French language, construction of French national iconography, and other social influences. Clearly today, on a supranational level, this type of iconography creation is not occurring except in select areas. Residents of Strasbourg are constantly reminded that their city represents a wider Europe, bigger than just France. This distinction is largely lost when there is little existing in the landscape to make real the mere idea of Europe. The construction of the *Palais de l'Europe* represents the growing desire and ability to reconstitute Europe into a physical thing, and place, more than an idea.

Le Jardin des Deux Rives

In 2002, a study was completed by the Communauté Urbaine de Strasbourg (CUS) regarding the feasibility and impact of constructing an international park through the joined forces of two towns, Strasbourg and Kehl. This park is scheduled to be called *Le Jardin des Deux Rives* (CUS, 2002). It will cross the Rhine River, the international border between Strasbourg and Kehl. The aim is not to provide an important transportation link between the two countries but instead to provide a link between the two cities through the creation of a foot bridge for weekend picnickers and bicyclists. This landscape alteration reflects the shift from national identities, denoted by the river as the boundary between states of Europe, to the emerging focus on local and supranational identity.

The conception of this new park vividly displays the attempts at creating symbolic connections in the landscape of two cities. This clearly visible park would offer a strong physical connection between these two cities ensuring the involvement of both populations through financial as well as moral support. While clearly this new park represents a city-to-city link, it is much more as well. Prior to 2002, Strasbourg and Kehl, indeed France and Germany by proxy, have maintained one international link between the two countries in this area, *le pont de l'Europe* [the Europe Bridge]. This construction would effectively double the international links between Germany and France following over 50 years of separation. But the growing interconnectedness of France and Germany, both economically and culturally, and the stresses on transport between the two countries is not to be cured by a foot bridge between the two countries; rather this landscape alteration is designed specifically to address the growing link between the two historically divergent cultures at the local level.

Le Jardin des Deux Rives is designed to provide a five km-long link between the residents of the two towns rather than the residents of the two countries in which these towns reside. In this sense, the link is a very different one than the connection offered by the Europe Bridge. Though discussed in the impact study in its relation to each village, it is clear that the construction of this foot bridge has implications beyond the local level as well. It points to something greater, Strasbourg's quest to be seen in an international context, or even a *Rhénane* (Rhineland) one, but assuredly not merely French or German. This construction of a park represents the construction of Strasbourg's international orientation and agenda, while remaining decidedly *Strasbourgoise*.

The park also offers the possibility for frequent day trips from just across the river. The cities, while separated by an international border, are united by proximity within the region. And because Strasbourg is not simply a French town, but has been influenced substantially by German culture, the city will be read differently by those "tourists" coming from nearby Kehl who plan on spending a day in Strasbourg, than by those coming from further away to vacation in Alsace or those coming to the city as part of the European Union's government. For those crossing the border, going to the next town over not only involves travel between two countries, but also a move into a hyper-articulated supra-national "European" landscape.

The nature of the construction, seen in the light of history, clearly has major symbolic implications. As Strasbourg moves into the twenty-first century and its role in European governance continues to grow, the landscape of Strasbourg will continue to represent the ideals of supranationalism combined with a strong sense of local identity. This park, crossing the international border of the Rhine River while remaining inside the local boundaries of Strasbourg and Kehl, represents how place becomes so important to landscape analysis. This construction of a practical, usable, local park to represent cross-national affiliation and reconciliation offers another example of the strong connection between the landscape and the conception of local and supranational place which is the hallmark of present-day Strasbourg.[1]

Place Name: Identity's Stamp on the Landscape

Labeling within landscape is important because labeling space makes it place. Place names reflect subjective qualities we have regarding landscape features or indeed the entire landscape itself. The "English" landscape is now used to label many similar looking landscapes throughout the world. Arenas, public forums, houses, buildings, and centers of worship are humanity's mark on the earth. The alteration of a landscape is often followed by the naming of place. Society's label for a piece of territory identifies it as something that has been carved from the wilderness; some*thing* that has become some*place*. The newly named place then enters discourse and lore, with attachments of meaning. In Strasbourg, the naming of place is made more complicated because many languages have been used throughout its history for this purpose. Therefore, using one particular language, in the absence of others, is in itself a powerful statement of allegiance and identity. The reclamation of place through name change by the citizens of Strasbourg is symbolic of the transition of society in this time of globalization and supranationalism.

Place name change is often the arbitrary delineation of place with value laden ideals. In 1940, with the control of Strasbourg reverting to Hitler's Germany, the new occupiers acted quickly to rename the existing landscape to reflect their ideals and version of history. In a map published in 1944 by the German government we see that *Bismarkplatz* regained its prominence as the center of power in Strasbourg, with Hitler *Platz*, secondary to the great unifier of Germany, taking the place of

---

1   Since the time that this research was undertaken the cross-border cooperation in the region has become even stronger with the creation of the Strasbourg-Ortenau Eurodistrict in October 2005.

Broglie (Plan der Stadt Strassburg 1944). Clearly the premise behind renaming, seemingly a vain and valueless act of transforming the landscape, actually reflects the ideas of permanence of a society. Hitler was not requiring the name change for a short stay in the city; instead his name was to remain in prominence "for a thousand years." Labeling the landscape is very often a first step in the reclamation of space and the reification of place.

Visitors often wonder aloud in Strasbourg why the stops for the tramline are announced in German and French. In fact, tramline stops such as *Langstross/Grand Rue* are announced in Alsatian and French (Figure 3.6). In contrast to the wholesale transformation of place name seen during the German occupation of Alsace, the current slow, methodical transformation of place names from French to Alsatian represents a more gradual and fundamental transformation of identity from French to Alsatian.

Figure 3.6 The Revival of the Alsatian Language in Central Strasbourg – Langstross, Grand'Rue

Alsatian, a dialect of High German, has been spoken in Alsace for some fifteen centuries (Vassberg 1993). Through the years of national struggle between Germany and France linguistic control was a method used to instill identity on place. In fact, the speaking of Alsatian or French during periods of German control was grounds for imprisonment. Following World War II, Alsatian speakers faced severe backlash as it was equated with the Nazi occupiers (Vassberg 1993). Suddenly the use of French with all the associated perceptions of civility and correctness was favored, and the push to learn, speak and label in French was strong. In 1946, only 66.4 per cent of Alsatians spoke French, while 80.5 per cent did so in 1963, and in 1979 census takers did not even ask, under the assumption that everyone spoke French (Vassberg 1993).

Understanding the efficacy of Frenchification following World War II, the recent transformation of place names in Strasbourg becomes more significant. Over sixty years since the end of World War II the Alsatian language has begun a public and private revival. The town center of Strasbourg, the ancient city center, has begun a process of place name change to previously used Alsatian names. Within the myriad of streets comprising the urban web of downtown Strasbourg, Alsatian can be found on many street corners.

What is important about this current resurgence of public Alsatian is the supranational environment in which it is occurring. With support from the European Union in the form of official language councils, even dialects in less well represented areas are likely to find new life. Important here is that Strasbourg society as a whole has decided to display Alsatian publicly, on street signs throughout the city. This reflects the reemergence of a regional or local ideal in the population previously hidden for cultural reasons. As Smith (1991) noted, once an identity has taken root it is nearly impossible to eradicate, no matter how often it is tried. This is the case with the Alsatian language in Strasbourg. While by no means as common as the use of French, the signs of Alsatian reemergence are beginning to show on the streets and buildings of Strasbourg's modern urban landscape.

Conclusion

In this age of globalization and time-space compression, the state is no longer the primary place in the social consciousness. As decentralization continues in Europe, locales are finding their own voice on the international stage. Strasbourg exists as a prominent place in both the imagination of Strasbourg society and the concrete and bronze urban landscape.

The urban landscape of Strasbourg, France is a rich palimpsest of ideas, beliefs, and ideologies. Urban design, building style, and statuary and symbolism are all attempts to create in the physical what only exists in the mental. How the town is designed, how the roads are oriented, what statues are constructed and preserved and what types of buildings are built represent society's need to make place real. Identity stems from place.

Identity, landscape, and place do not exist in isolation, unrelated to one another. Instead, each element is directly related to the other. The interaction of landscape, place, and identity is the nexus of this investigation, and it is what the tourist confronts. Identity, as it is constructed in the modern world, hinges on the conception of place and the reification of place in the landscape. Each individual existing in society derives some of themselves from the interaction of landscape and place. On a larger scale, societies as a whole derive their identity from this complex interaction of landscape and place. The identity of a people is intimately related to the places they have created through altering their landscapes.

In this age of globalization and supranationalism the process of "individuation" (Harvey 1989, 302), or the process of carving out a niche in the world where the group is different from "others," becomes more important. Societies are reclaiming their spaces, creating new places, and their identities are being altered as a result.

In Europe, for example, old conceptions of place are once again becoming more important as a result of the fading borders of the states. Strasbourg is once again becoming a local place while it simultaneously evolves into a window to a wider Europe. Strasbourg's location astride the traditional dividing line between French and German cultures serves as a powerful symbol for Franco-German reconciliation. No longer is Strasbourg a contested region between two powerful states, but a location representative of a wider, pan-European ideal. The transition of identity from national/regional to local/supranational in Strasbourg from 1870–1970 to 1970–2002 represents a wider phenomenon occurring in Europe. The locales of Europe, as well as Europe itself, are becoming more important for identity formation. States and regions are no longer the primary foundation for identity.

Strasbourg represents how a place can help to create a distinct identity for both local citizens and the entire European population. Residents of Strasbourg can simultaneously infuse their identity into the city and draw their identity from it. Residents of the growing European Union can also use the symbol of European unity—Strasbourg—to develop their own European international identity. Ultimately, the reification of place in the urban landscape of Strasbourg represents a growth of a local and supranational identity in the heart of Europe which compliments, and in many ways supersedes, the importance of national and regional identity which predominated at earlier periods in its history.

The identity–place–landscape nexus is an important one for tourism since it is fundamentally the nexus that the tourist experiences. However, the shifting terrain of place and its relationship to identity highlights the difficulty of the task facing the tourist to Strasbourg, and for that matter, any other place. The visitor confronts a landscape at one point in time. Yet the tourism landscape is comprised of multiple facets or layers that have been built up, eroded and rebuilt over a considerable period of time.

Strasbourg offers a unique site, a landscape rich in the historical and cultural patrimony of its varied past—from Celtic, Roman, Alsatian, German, and French influences—and a landscape rich in the possibilities of creating a supranational identity. The pan-European landscape is in some cases more deliberate in its creation than the Alsatian. For example, it has been intentionally designed by professionals rather than emerging from the vernacular practices of the local community. The creation of this European quarter is an example of European identity put into concrete form adding yet another layer of meaning onto the city. Strasbourg as a "European" capital presents itself not only to those tourists who travel from afar for leisure, but also to those "tourists" who come to work with the invasion of the European bureaucracy. Additionally, there are those who come to *Le Jardin des Deux Rives* from the German town of Kehl. The landscape that each of these various tourists encounter is the same on the ground but the meaning depends, it can be said, upon the geographic scale of reference. The typical tourism landscape, like that of Strasbourg, is by turns local, regional, national, and supranational; so it is hardly surprising that this landscape, like the statue of LeClerc, is a repository of multiple identities, sometimes ironic, sometimes at variance, and never facile.

Chapter 4

Landscape Change and Regional Identity in the Copper Canyon Region

Yamir González-Vélez

The Sierra Tarahumara is still a pastoral reality, a place where there are real Indians and genuine Vaqueros…is very unlike the fantasy world of six-guns, wild women, "redskins", and whiskey that dances in the head of suburban cowboys in Laramie and Cheyenne. [Raat and Janeček 1996, 9]

Landscape and Identity

Landscapes are a critical source of social orientation and collective identity (Kaufman and Zimmer 1998). National identity is characterized by being both cultural and ideological, therefore it needs to be imagined, performed, and witnessed in order to be recognized as such (Poulter 2004). Although national identity does not necessarily have to be associated with a state, it inevitably is associated with territory.

Traditionally, national landscapes have served as both the source and the product of national identities. National landscapes are defined as "the landscape or set of landscapes that represents and identi[fies] the values and essence of the nation in the collective imagination" (Nogue and Vincente 2004, 117). National landscapes can be seen as the transformation of ideas into palpable realities, creating or strengthening the ties through which particular factions of the population can identify themselves as a people (Kaufman and Zimmer 1998).

However, landscapes that carry enough symbolic meaning to be considered national landscapes are by no means unproblematic. Indeed, most such landscapes are the focus of some, if not considerable, conflict. Deeply symbolic landscapes often are the focus of contestation between those that wish to preserve it, those that wish to use it as a means of livelihood by extracting its special qualities (whether that be mineral wealth or scenic beauty) and the landscapes' original inhabitants.

It is also important to recall that the designation of landscapes as "special" or culturally distinctive in some symbolic way occurs at a variety of geographical scales ranging from the urban to regional, to national, and to global (e.g. World Heritage Sites). This chapter evaluates the existence of a regional landscape in the Copper Canyon region of Mexico by exploring the changes in the landscape of the town of Creel during the last century. More specifically, this chapter examines the way in which different groups imagine, perform, and view the landscape of the region. In this area three major actors (groups) can be identified, the *Rarámuri*, the *Mestizos*, and the "other."

For the *Rarámuri*, the canyons have been natural barriers that protect them from outside influence, and their main source of livelihood because they use the canyons for hunting and gathering, agricultural production, and limited but important livestock production. Most importantly, however, it is their home, and holds spiritual meaning as the place in which the souls of their ancestors come to dwell.

Like the *Rarámuri*, the *Mestizos* are at home in the area, though they do not see the canyons as protective shields or spiritual places. For the *Mestizos*, the canyons are a source of income, through the exploitation of the natural resources that it contains (timber, medicinal plants, etc.) and through the exploitation of its scenic beauty through tourism. They manage and control most of the economic activity in the area.

The third group to be considered is a collection of entities that have sundry interests in the region (ranging from economic to religious) but do not necessarily have historical or ancestral ties to the land. These people will be referred to as "the other." Within this category, the government and tourists stand out. Both groups seek to exploit the region for its economic potential and both, in opposition to the *Rarámuri* and the *Mestizos*, are outsiders to the region. Most importantly, these groups engage in promoting a kind of development that relies upon the scenery and upon the exotic. The canyons are seen as picture perfect experiences that, if framed well, can be used to jumpstart an economy which has been experiencing a recession since the 1980s. For this group, the resources of the region are transformed into commodities, topography is turned into scenery, and people into spectacles.

Methodology

In order to explore the changes of the Copper Canyon, a landscape methodology was utilized in this study. Techniques include detailed fieldwork, participant observation and the use of historical analysis in creating a case study. Landscape studies entail a painstaking attention to detail in observing the physical and cultural features of the site and recording observations through note taking and through photography. Given the nature of the work and the complexity that is involved in landscape studies, the case study method is a useful research strategy for facilitating the way in which both qualitative and quantitative data are utilized (Yin 1981). While the region itself is the subject of this study, research focused on the town of Creel, which has traditionally served as the window to the area, the town of El Divisidero, and the infrastructure which provides access to the Copper Canyon, including the *Chihuahua al Pacífico* railroad and the extensive road infrastructure which has been built up in the last century.

Field work was undertaken in January and August 2004. On both occasions the train and the roads were utilized for transport as well as objects of study. Participant observation was undertaken on the train trips, which included a three hour journey between Creel and El Divisidero and a six hour ride from Chihuahua to Creel. The newly paved road from El Paso, which has allowed easier access to the region for tourists and the movement of natural resources out of the area, was also traveled. Extensive hiking was undertaken in the region of the canyons. The towns of Creel and El Divisidero, including the tourist sites at both, were explored and changes between the two time periods were noted. In addition, during the second field trip, the downtown[1] area of Creel was surveyed and mapped. Informal interviews were conducted with local authorities and representatives of the tourist industry in the towns and in Chihuahua informal interviews were conducted with government officials, including the director of *Gobierno del Estado de Chihuahua, Coordinación Estatal de la Tarahumara* (the office of Tarahumara affairs). Detailed maps of the region and demographic information were obtained from state and federal agencies during this visit.

Because the landscape is ever evolving, as Sauer (1925) noted in his statement on landscape methodology, it is important to utilize a wide array of sources in addition to the physical entity of the landscape itself in order to get the full picture. To this end, historical documents, newspaper articles, government reports, and guides for tourists, amongst other sources, were used to asses the status of the landscape of the region and the rate of change which has been affected by various actors in the region. Historical maps in juxtaposition with "popular"[2] maps were used to determine accessibility to the region and road construction. A base map of Creel was obtained from INEGI[3] and with the help of Arc GIS, two detailed maps were generated. The first map (see Map 4.1) represented the town as surveyed by the local government in 1993, the second map (see Map 4.2) was generated with the data obtain during the second field trip to the area.

The Copper Canyon Region

The Copper Canyon region is located in the state of Chihuahua in northern Mexico. Situated within the Sierra Madre Occidental mountainous system, the region is characterized by agricultural production, ranching, and timber harvesting. The diversity of climate and soils allows for the establishment of different types of crops, which has made the region one of the largest biologically diverse areas in Mexico (SAGARPA 2001). It also contains about two-thirds of the standing timber in Mexico (TED 1996).

The geographical area which constitutes the Copper Canyon region is defined differently by each one of the groups that have an interest in the region, ranging from the perspective of the indigenous populations to that of the federal government. Maps that accurately define the area are difficult to find because the area has been transformed into an important tourist region and maps often emphasize particular locations, such as tourist attractions or hotels, as well as the areas surrounding the *Chihuahua al Pacífico* Railroad. Topographically, the Copper Canyon region is comprised of a system of canyons rather than a single canyon (see Figure 4.1).

---

1   Given the size of the town and its population the term "downtown" is used to make reference to the main streets that comprise the town, this being *Avenida Gran Vision* and *Avenida Francisco Villa*.

2   The term "popular" is used to make reference to maps provided to tourist by local vendors.

3   Instituto Nacional de Estadística Geografía e Informática de México.

The area is made up of four gorges over 1,500m in depth: the Upper Canyon *Barranca del Cobre*, the Lower Canyon *Barranca del Urique*, the southern gorge *Sinforosa Canyon*, and in between them, the *Batopilas Canyon*. The term "Copper Canyon" comes from the copper/green colored lichen that clings to the canyon walls (Cummings 1994). While the physical landscape of the region is apparent to all, the three groups of actors in the region have imagined, performed and witnessed the physical landscape in various ways with concomitant implications for the landscape itself. Together these groups have defined the identity of the Copper Canyon region though a collective imagination of the area is lacking because each group has a different perspective (Meinig 1979).

Figure 4.1 Copper Canyon Wall

The *Rarámuri*

The Indians of the *Sierra Tarahumara* are the largest group of Indians north of Mexico City apart from the Navajo (Kennedy 1978). Although the *Sierra Madre* is the home of several indigenous groups (*Pima, Wario, Odame,* and *Rarámuri*), most scholars tend to refer to the population of the region by the name of the largest group, the *Rarámuri* (Cassel 1969; Kennedy 1978; Zingg 2001). Approximately 50,000 *Rarámuri* Indians live in the *Sierra Tarahumara* (Raat and Janeček 1996). Their territory used to be much larger, but through processes of displacement and with the "mestization" of the indigenous population, their territory is now restricted, for the most part, to the most inaccessible areas of the western *Sierra Madre* mountain chain, i.e. the Copper Canyon (Kennedy 1978). Little is known about the *Rarámuri* culture and lifestyle before and during the initial period of colonization, although letters and documents from missionaries and other officials dating to this period of early contact have been preserved in northwestern Mexico (Kennedy 1978). The first recorded contact with the *Rarámuri* Indians was made in 1607 by a Jesuit missionary from Spain, and between 1607 and 1611, meaningful interaction with the *Rarámuri* began (Cassel 1969; Kennedy 1978).

For most of the colonial period, however, the government and the missionaries had little interaction with the *Rarámuri* as the region was considered inhospitable with few apparent resources (Kennedy 1978). As the missionaries began slowly moving into the area, mineral wealth was discovered and mines were opened in *Parral* and other parts of the region (Fontana and Schaefer 1997), precipitating a boom that resulted in the displacement of the indigenous population. Considerable amounts of Indian land were taken over during the seventeenth and the eighteenth century by the new settlers (Kennedy 1978). The Indians of the *Tarahumara* tried to protect themselves from outsiders through isolation, and a pattern of avoidance and withdrawal into the mountains became established (Kennedy 1978). Towards the end of the nineteenth century, both church and state began to take a renewed interest in the *Tarahumara* region (Kennedy 1978; Fontana and Schaefer 1997), resulting in the establishment of the town of Creel, which corresponded with the arrival of the *Chihuahua al Pacífico* Railroad, or *Chepe*, into Mexico (Cummings 1994; Kennedy 1978).

The 1910 Mexican Revolution drastically changed the *Rarámuri*'s right to the land. Between 1910 and 1920, the economic use as well as the ownership of the land in the Sierras changed (Anderson 1994). During this time period, the supply routes from the mines were cut, no food could go in nor could gold or silver go out (Anderson 1994). Communal ownership of the land, a widely known practice by indigenous peoples throughout Latin America before the colonial era, returned in the form of the *ejido*. By 1917, a new constitution was approved in which the *ejido* became the basic legal form of land ownership in the Mexican country side. Agricultural land was expropriated from large landowners and redistributed to communal farmers (MacLachlan and Beezley 1994). The *Rarámuri* gained legal rights to the land that they occupied; parcels were held individually while the governance of land use and common land such as forests, were decided by *ejidarios*, which included every male head of a family (Anderson 1994). Although the *Rarámuri* had lost the valley in which the town of Creel was situated, they still occupied the surrounding hills and valleys, and outside encroachment came to a temporary standstill thanks to the new land rights law (Anderson 1994).

During the twentieth century, there were few outside attempts to centralize the disparate segments of the *Rarámuri* population through missions or government action, and they managed to remain a unique population with traditional practices for most of the modern era (Kennedy 1978; Fontana and Schaefer 1997). In the past, the animals found in the region were hunted for food, but a number of these, including big horn sheep, the gray wolf, the mule, white-tailed deer, the Mexican grizzly, the black bear, the mountain lion, and the jaguar, are now extinct or nearly so (Raat and Janeček 1996). These changes in the health and numbers of wild species have led to changes to the *Rarámuri* lifestyle. Livestock breeding, mainly goats and cows, is now the primary activity, but many engage in some agriculture as well. The principal crop is maize (*sunúku*), although they also cultivate beans, squash, wheat, chilies, potatoes, peaches, oranges, and tobacco (Raat and Janeček 1996; Kennedy 1978). Herds act as a symbol of wealth and prestige (Kennedy 1978). Goats are especially prized for the fertilizer they produce, and herding is important ecologically because it imposes rigorous requirements on the *Rarámuri* way of life with respect to the landscape of the area. Herds are generally small to take advantage of the small patches of grazing land which can be found amongst the steep hills and cliffs. Animals are also seen as a form of savings, since in times of severe drought livestock serve as another source for subsistence (Kennedy 1978).

The culture of the *Rarámuri* is intrinsically bound to the physical environment of the *Sierra* and the horticultural and hunting practices which have developed in response to the specific nature of the landscape (Kennedy 1978). But as will be seen below, the changes in the landscape brought about by the *Mestizo* population and "others" such as the federal government has resulted in the lifestyle and the culture of the *Rarámuri* being modified.

Mestizos

> ... the invaders see and think differently. To quench their thirst they behave like beasts of prey; they cut down the forest and open the Earth-without respect and with violence. They forget that the Earth is a living body. (Anonymous Rarámuri in Raat and Janeček 1996, 15-6)

In the Copper Canyon region, the first attempts of colonization were made by Jesuits in the seventeenth century (Kennedy 1978). While the missionaries were slowly moving into the *Tarahumara* region, mines were discovered in *Parral* and other parts of the region (Fontana and Schaefer 1997), precipitating a migration boom that resulted in the displacement of the indigenous group and the creation of a *Mestizo* population in the area. As in the rest of the Americas, the *Mestizo* population in this area is the product of a history of colonization and racial mixing. Because of its rugged terrain and isolation, the building and maintenance of formal settlements proved to be a challenge; during the colonial and early revolutionary periods most of those who came to the area came as the result of the establishment of mining camps and logging concessions. Much Indian land was usurped during the seventeenth and eighteenth centuries by the new settlers though accounts of formal settlements can only be found at the end of the nineteenth century (Kennedy 1978).

After the era of Spanish colonization, a new period in Mexican history opened with the Mexican revolution. Although new laws were approved to protect the native population, these laws in many cases favored the *Mestizo* population. The right of the "Mexican" to own land was recognized but only if they were "civilized" citizens, with the result that the *Mestizo* population was privileged over that of the Indians (Kennedy 1978).

Two other events affected the region, the arrival of the *Chepe* railroad and the founding of the town of Creel. The idea of the *Chihuahua–Topolobampo* route, which included the Copper Canyon region, was conceived by engineer Albert Kinsey Owens whose joint Mexican and American company, founded in 1863, was granted the contract to build the railway (Zingg 2001; Cummings 1994). Because he failed to acquire enough investors, his contracts were taken over by Foster Higgins and his Sierra Madre and Pacific Railway Company (Cummings 1994; Kerr and Donovan 1968). By 1898, the Higgins Company was responsible for 259 kilometers of rail lines between *Ciudad Juarez* and *Casas Grandes* though it had not yet reached what would become Creel (Raat and Janeček 1996; Wampler 1969).

Afterwards *Mestizo* entrepreneur Enrique C. Creel[4] and his Kansas City, Mexico, and Orient Railroad Company took over the project (Cummings 1994). In 1906, Creel established the eponymous colony of Creel in the Copper Canyon region (Cummings 1994; Kennedy 1978). The idea behind the colony was to assist the indigenous population. In theory, the new "colony" would be composed of 75 per cent *Tarahumara* and 25 per cent Mexican, but the settlement of Creel was more beneficial to Mexican families than to *Tarahumaras* (Fontana and Schaefer 1997; Kennedy 1978). After many economic and political difficulties, the Kansas City Railroad, what would later become the *Chihuahua al Pacífico*, arrived in Creel in 1907 (Raat and Janeček 1996). The arrival of the railroad revitalized the colonial mining settlements, leading to the rise of a new class of regional elites (Knight 1986). The communities of the area suffered under the control of these regional elites, whose power extended to the political realm (Knight 1986). The town of Creel served as the base of operation for the *Mestizo* population and the access point to the rest of the Copper Canyon region. It also became the shipping point for minerals and timber out of the area. In the eyes of this new wave of settlers, the landscape of the *Tarahumara* was a source for generating capital by exploiting the natural resources of the region.

"Other"

Although construction of the *Chihuahua al Pacífico* Railroad had begun in the early 1900s, a number of factors impeded its completion including the difficulty of the terrain, the Mexican Revolution, mismanagement, and lack of money (Lister and

---

4   In the Porfiriato era, in Mexico Enrique C. Creel served as the interim, and later, governor of the Sate of Chihuahua, he also served as Ambassador of Mexico in the United Sates and Minister of Foreign Affairs in Mexico.

Lister 1966). It was only in 1953 that the federal government announced its intention to complete the route (Kerr and Donovan 1968). After 18 years of tunneling and laying tracks, the gap was closed in November of 1961 (Cummings 1994). Parallel with the efforts to complete the railroad line, construction was begun on new roads in the region as a response to the needs of the timber and the mining industry. Prior to the completion of the rail line, between 1945 and 1948 a one-lane road was cut from Creel to La Bufa, a distance of approximately 100 kilometers so that heavy trucks could move the rich copper concentrate from the Carmen mine to the railhead at Creel. The road was continued another 32 km between La Bufa and Batopilas, but despite the new roads, accessibility continued to be a problem.

The completion of the *Chihuahua al Pacífico* Railroad, however, marked the opening of Creel to the influence of outsiders, among them, the federal government and tourists, and has resulted in major changes to the landscape. In the 1980s, Creel was considered Mexico's leading producer of wood, plywood, cellulose, veneer logs, ponderosa pine, but production had decreased by 1991 (Raat and Janeček 1996). By the end of the 1980s, Mexico was one of many countries suffering from the effects of a global economic recession, and in 1989 the World Bank loaned Mexico $45.5 million for a logging and forest-management project (Mardon and Borowitz 1990 in TED 1996). The revenue from timber harvesting was intended to help ease the economic crisis by helping Mexico pay its foreign debts as well as reducing its dependence on imported paper pulp. The plan was to log more than four billion board feet of lumber from 20 million acres of forest over six and a half years (TED 1996). As part of the project, then Mexican President Carlos Salinas de Gortari and his government devised a plan to displace the indigenous population from the Copper Canyon region. His plan failed when, in October of 1991, the Act of Dislocation[5] was made public. Facing international pressure, the act was never pursued (Raat and Janeček 1996; TED 1996). In 1993, gold mining interests pushed for further road construction and a new road was cut from Batopilas through Santevó to San Ignacio, close to the Sinaloa border. This new road not only provided a transportation and commercial link for legitimate activities, but also increased the activity of drug traffickers (Raat and Janeček 1996).

In spite of the government's failure in expelling the indigenous population, it was not ready to abandon the idea of exploiting the resources of the region, and by 1994 a new paved road, Highway 16, from Chihuahua to Creel, was finished (Cummings 1994). Concomitantly in 1994, President Salinas signed the North America Free Trade Agreement (NAFTA), which led to an increase in foreign investment not only at the national level but at the local level as well. Changes in the region started with the re-evaluation of property rights. To comply with NAFTA, the Mexican government had to make changes in the Mexican Constitution regarding the ownership of *ejido* lands. Indigenous people became more vulnerable as they were pressured into selling their land to either the government or private investors (see Map 4.1).

---

5   The 1991 Act of Dislocation was an attempt to privatize some of the communal lands of the *Ejido* of San Ignacio, in order to allow for the development of private enterprises in the land adjacent to the Arareco Lake near Creel (Raat and Janeček 1996).

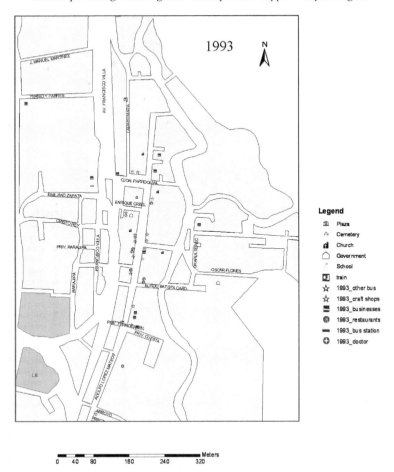

Map 4.1 Creel, Mexico in 1993

Two years later, President Ernesto Zedillo and his administration backed a new plan for the development of the region. *El Plan Maestro de Desarrollo de las Barrancas del Cobre* was intended to transform the region into a center of national and international tourism. Prior to the arrival of the train in 1961, "most tourists, foreign or Mexican, had never laid eyes on a Tarahumara"—as the *Rarámuri* are referred to by outsiders (Raat and Janeček 1996, 140). However, in the ensuing years the region developed into a major tourism destination. Besides opening the region to intensified extraction of resources, the train also became a major tourist attraction. As it received media attention in newspapers and magazines, the ride increased in popularity, and its inclusion in guide books served to boost the tourist industry in the region, propelling the establishment of hotels and organized tours (Anderson 1994). It was not just the landscape which drew visitors, but increasingly the *Rarámuri* became a tourist attraction as well (Anderson 1994).

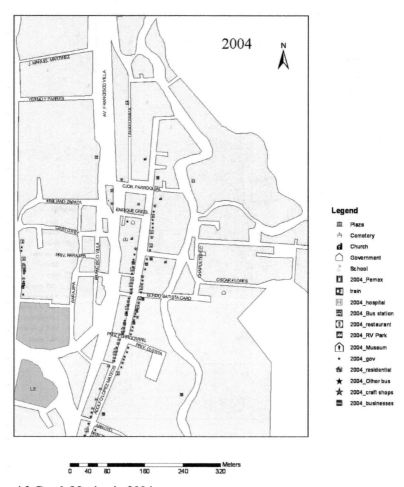

Map 4.2 Creel, Mexico in 2004

The federal government's plan called for an initial investment of $91 million aimed at improving the infrastructure of the region (Perez 1997). That year, Mexican President Ernesto Zedillo visited the town of Guachochi and inaugurated the newest section of a paved road that connected the region (the new section connects Chihuahua to Guachochi) via paved road to the West Coast of Mexico. During the ceremony, Zedillo handed over a check to finance the last stage of the road project, to connect Creel to El Divisadero via paved road, and in his speech he promoted Creel as a tourist destination. To achieve this goal, the plan proposed the strengthening of infrastructure (roadways, electricity, telephone, and water treatment), and the establishment of at least 900 hotel rooms, 300 camp sites, and 200 spaces for recreational vehicles.

The first phase of the plan was put into motion fairly quickly with the local government expropriating most of the land in Creel. Opposition to the plan, however, forced the government to re-sell the land to prior tenants above market price, but as land value and property taxes rose in relation to development locals were forced to sell (Perez 1997). The emphasis on tourism has meant that national and international forces have continued to reshape the landscape of Creel. In 1998, a new Best Western Hotel was built, and in 2000 the completion of the paved road to El Divisadero decreased travel time by bus and private transportation dramatically. It now became possible to drive on a two lane highway from El Paso to Creel in approximately eight hours. In 2002, the RV campsite was finished and since 2003, the region has begun to be marketed as an eco-adventure and extreme sports destination (see Map 4.2 and Figure 4.2).

The second phase of *El Plan Maestro de Desarrollo de las Barrancas del Cobre*, which began after 2000, has come at a slower pace; international human rights groups have pressured the government to protect the indigenous population of the area, causing the government to reconsider its plans, although development in the area is still heavily promoted. In the late 1990s, the government proposed to designate the region a UNESCO World Heritage site but did not follow through with the process (Osorio 2002). Instead, foreign investments have left a mark on the town—Creel is turning into a strip mall (Perez 1997).

Figure 4.2 Hotel Villa Mexicana Lodging Resort and RV Park

The indigenous population of the area is still suffering the consequences of the development plan as well as the effects of resource extraction through timber and mining. In terms of natural resources, the rapid depletion of the timber reserves in the area and the over-cutting of the forests have caused high levels of deforestation as well as the extinction of the imperial woodpecker (the largest woodpecker on Earth), and the endangerment of fauna such as the Mexican gray wolf, jaguar, and thick-billed parrot (TED 1996). The way of life of the *Rarámuri* has been affected as they are forced out of a territory that has historically been used for hunting and gathering and agriculture. With their livelihoods in danger they have sought to develop new strategies for survival, turning to the production of craft-based artifacts to be sold to tourists. Both women and men work on the production of baskets, pottery, drums, blankets, pine-bark dolls, wood carvings, violins, flutes, necklaces, and belts. Some are authentic *Rarámuri* household items, while others are made just for tourists (Anderson 1994). The indigenous population in Creel has become part of the scenery as their rights and needs have not been considered by developers, including the government. The increased accessibility of the region has transformed not only the physical landscape but the perceptions of the region as well. It is not surprising that when combined with the scenery provided by the train and the indigenous peoples engaging in craft production, the Copper Canyon region gives the tourist an illusion of an authentic and exotic place.

Conclusion

The development of a transportation infrastructure, such as constructions of new paved roads and the completion of the railroad under the influence of "others," outsiders to the area, has significantly reconfigured the landscape of the Copper Canyon region. Since the inauguration of the new railway line in 1961, notions of time and space have changed in the region as well. The new railroad not only physically connects Creel with Los Mochis and the West Coast of Mexico (Raat and Janeček 1996), but it also connects its resources, allowing for an exchange of goods and services, resulting in an expansion of the economy.

The railroad serves as an input as well as an output for the resources of the region, in that materials such as timber, agricultural products, as well as construction materials, and more importantly, people can finally be moved across the mountains in an efficient way (Anderson 1994). The new transportation facilities have made the area more accessible, but the increasing influx of outsiders has had a profound impact upon the landscape and the way of life of its residents, indigenous, and *Mestizo*. Ranchers have already taken the best land; drug traffickers have infiltrated the area attracting the police and the army (Raat and Janeček 1996). Currently, most of the territory is unsuitable as farm land and scarcely capable of providing enough agricultural produce for local consumption. Mining activities are declining and local timber and other forest resources have been over-exploited, affecting the livelihoods of the local *Mestizo* population (Perez 1997).

In the eyes of the federal government, tourism has become the preferred solution to the economic hardship that this region is undergoing. The development plans have resulted in even more construction of roads and other infrastructure—such as the building of international hotel chains and RV parks—to support the flow of national and international tourists. However, in their attempt to solve an economic crisis, the government is causing a cultural one. The *Mestizo* residents of Creel have transformed their front rooms into stores and restaurants while others began renting bedrooms and advertised their homes as guest houses while the *Rarámuri* have become cultural attractions for tourists, adopting craft production, sometimes "authentic" and sometimes spurious, as a way to make a living (Anderson 1994).

The lack of a collective imagination among the three players in the region and the existence of different value systems, practices, and perspectives on the land have resulted in a landscape which is seen not just differently, but in mutually incompatible ways. If national landscapes are defined as "the landscape or set of landscapes that represent and identify the values and essence of the nation in a collective imagination" (Kaufman and Zimmer 1998), then after evaluating the case of the Copper Canyon region and the town of Creel there is a lack of "territorial ideology" (Nogue and Vicente 2004). Globalization has reached *La Sierra*, it has managed to precipitate the acculturation of indigenous people, it has transformed the canyons from sheltering structures to picture perfect realities, and it has contributed to bringing "real Indians and genuine Vaqueros" back from the past in order to create a new type of collective memory.

Chapter 5

Mauritian Landscapes of Culture, Identity, and Tourism

Anne K. Soper

Introduction

The government of Mauritius has turned to tourism development in recent years as a means of diversifying the economy of this small, peripheral island in the Indian Ocean. Increasing pressure is being placed on the local tourism sector as other main pillars of the economy are in decline and the tourism industry itself is now in a position where it too needs to diversify. Sun, sea, and sand tourism, for which Mauritius is known, will continue to be the major focus of the island's tourist industry. However, recently, traditional tourism is being supplemented by the introduction of local history, heritage and culture as tourism products for consumption by both international and domestic tourists. This chapter examines some of the places and events chosen for cultural heritage tourism, and how, in the process of becoming part of the tourism landscape, these places may aid in the construction of a collective Mauritian memory. The chapter will unfold with a brief history of Mauritius, detailing the socio-cultural mix of people who call themselves Mauritian. It will then identify some of the issues confronting Mauritius as it struggles to develop a more concise national identity, and finally, the chapter will examine how expanding the tourism product to include more aspects of Mauritius' cultural heritage is aiding in the strengthening of an integrative Mauritian nation.

Methods

The data in this chapter is derived from nearly a year of field research in Mauritius. During that time, many cultural heritage sites, whether officially designated or not, were visited. Structured and unstructured interviews were conducted with the stewards of the sites, academics at the University of Mauritius, Government officials, NGO leaders, tour operators and Mauritian citizens who were not directly involved in the tourism industry. Much of the information pertaining to *Aapravasi Ghat* was gathered while attending meetings of the Aapravasi Ghat Trust Fund in the role of observer-participant. The quantitative data provided here reflects a very small portion of a lengthy structured questionnaire administered to 1,198 Mauritian citizens that was analyzed using cross-tabulation and chi-square to test for significance. Historical documents and other secondary sources were viewed at the Mauritius National Archives and other libraries, the Mahatma Gandhi Institute, museums and various online sources.

The Making of Mauritius

Located in the southwestern Indian Ocean, Mauritius is a small, tropical island approximately 500 miles east of Madagascar. Although it is volcanic in origin and has a mountainous interior, the island boasts over one hundred miles of white sandy beaches and clear blue lagoons that are protected by a coral reef that surrounds the island. All these attributes have led tourists and locals alike to refer to Mauritius as "paradise." Aside from its "paradise" status, Mauritius is also considered to be one of Africa's "miracles" as a result of its unprecedented economic development and growth.

Discovery by the Portuguese nearly 500 years ago, the island was a stopping point for seafarers crossing the Indian Ocean affording them a place where they could rest and repair their vessels before continuing on their voyages. It was the Dutch who first colonized Mauritius with a short-lived first attempt from 1638–1658 and again from 1664–1710. The military significance and strategic location of the island was identified by the French who began their occupation of the island in 1725, but lost the island to the British in 1810. British rule ended in 1968 when Mauritius gained its independence from Britain. The colonial legacy lives on in Mauritius and the country has functioned as a British-style democracy dominated by a Hindu majority. Since the time of independence Mauritius has grown and changed tremendously. It has gone from a less developed country dominated by a plantation style sugar monoculture to a tourism paradise, an export oriented textile manufacturer, and is now striving to become a cyber-island which would further integrate it with the rest of the world.

All of these roles are being pursued concurrently with a fair amount of ebb and flow between economic sectors. Sugar is no longer king and textile manufacturing is witnessing a downturn as the cost of labor rises. The cybercity is in its embryonic stage and heavily connected to India. Nearly every section of the Mauritian economy is searching for new and different ways to do business. The sugar industry is exploring markets for expensive specialty sugars, ways to use *bagasse* (the crushed remains of cane plants) for everything from power plant fuel to cloth, and has opened a museum that tells the story of sugar in Mauritius. Tourism is another Mauritian industry that is marked for change, not because it is lagging, but to encourage greater growth by adding dimensions. With a worldwide reputation as an island paradise, Mauritius has no desire to overhaul its main offering—the beach experience. However, it sees value in utilizing Mauritian history, heritage, and culture as new avenues for tourism development.

Mauritian Identity

Mauritius is young, as countries go, but because of its colonial past, it has an abundance of official documentation detailing histories of people, places, and events.

Despite this official documentation, as noted Mauritian historian Benoit explains, some lapses and revisionist history, particularly with regard to slavery and other contentious and discomforting events, have occurred (Gordon-Gentil 2003). As a result, issues of heritage and culture in Mauritius are not straightforward. There are no indigenous Mauritians. The island was populated by people originating from a multitude of places, who followed an array of religions and traditions, and spoke their ancestral languages. This multi-ethnic gathering was a direct result of historical circumstances which brought European colonialism and a sugar plantation system to the island. With no indigenous labor available, African slaves and Indian indentured servants were brought to Mauritius, followed by traders and merchants from China.

Present-day Mauritius consists of a population of just over one million, and no less than five major ethnic groups: Indians, who are separated into two main religious groups—Hindus and Muslims; Creoles—those of African descent and mixed race persons; Franco-Mauritians who are white Christians; and Chinese, who are mostly Christian as well. Indians make up the majority of the Mauritian population, but cannot be viewed as a homogeneous group since there are several ethnic subcategories such as the Marathis, Tamils, and Telegus, each claiming their own cultural identity. Likewise, Muslims are self categorized according to sect. Despite the fact that both the Hindus and Muslims originated from India, the groups are very distinct due to a long-standing socio-political cleavage which reflects geopolitical patterns in India dating back centuries. The Creole population is heterogeneous by definition. Franco-Mauritians and Sino-Mauritians are the most homogenous groups found on the island, but are minorities representing a very small percent of the overall population (for a complete ethnography of Mauritius see Eriksen 1988).

In Mauritian society ethnic groups and sub-groups are called "communities" and community membership is an integral part of one's individual or personal identity. Being a member of one community does not prevent a person from participating in the rituals or practices of another group. For example, it is not unheard of for Hindus to lay offerings of flowers at the feet of Christian saints. Another case may be the taxi driver who has symbols from several religions enshrined on his dashboard "just in case." Mauritian cuisine as well can be described as an amalgamation of ethnic foods, each enjoyed as equally as the next. Chinese noodles, or *mine*, are enjoyed by many Mauritians regardless of ethnic background, as well as Indian curries, and *biryani* (a rice dish). In short, Mauritians practice shifting identities and feel comfortable moving between identities. Circumstances will sometimes dictate when it is expected to identify with one's community and when it is acceptable to shift. Eriksen (1997) has identified several facets of life in Mauritius including language, trade unions (tourism workers, in particular), nationally produced television programming, enjoyment of the beaches, and universal movements such as women's rights and environmental protection as being "irrelevant" to ethnicity. On the other hand, mixed marriages, league sports (specifically soccer teams of the 1980s), and youth centers are "seriously challenged" by ethnic boundaries. Still, the lines are not always clearly drawn as Benoit discovered when doing scholarly research on a community other than the Creole community—the one to which he is, due to his ethnicity, a member by default. He reports having been forbidden to deliver a research-based presentation on the Sino-Mauritian diaspora at an international conference and says

"It seems that in Mauritius, one should only talk about his community" (translation from Gordon-Gentil 2003, 8). Although this experience occurred several years prior to the 2003 newspaper interview, it demonstrates that difficulties can be encountered when crossing between identities.

In recent years, the government has shown support for the separate community identities by officially recognizing, establishing, and funding cultural centers. This move is not without controversy as it appears to be endorsing a move away from an inclusive Mauritian national identity. Promoting the slogan "Unity in Diversity" as a principle to live by, the government has demonstrated a desire for all Mauritians to respect each individual culture while forging a cohesive national identity. The thinking here is that the whole is only as strong as each of its component parts. Therefore, encouraging the establishment of centers focused on the preservation, promotion, and practices of cultural groups will strengthen its members. The more secure people feel in their individual cultural beliefs, the stronger the Mauritian nation as a whole will be. Some cultural centers in Mauritius have a long history— *Le Centre Culturel Français* and the British Council are two such centers. Likewise, the Chinese Cultural Center and the Indira Gandhi Centre for Indian Culture are well-established organizations. All of these centers are funded externally by the home country. Other centers are homegrown affairs, such as the Marathi Cultural Centre, Telegu Cultural Centre, Tamil Cultural Centre, the Islamic Cultural Center, the Nelson Mandela Centre for African Culture, and the most recent addition, the Mauritian Cultural Centre. Arguments have erupted over many questions surrounding the designation of cultural centers such as the amount of government subsidies given to each center (Hills 2002), whether a location for the center is to be specially designated and built with government support or rented by the governing members of the center, and whether the establishment of all these centers will be beneficial to the nation. Presently, the centers are not part of the tourism landscape and do not benefit the nation through tourism revenue. There is a great deal of concern that the centers will divide Mauritians into ethnic enclaves instead of bringing them together in the spirit of Mauritianness.

Identity and Tourism

There is, as yet no fixed Mauritian national identity or agreed upon definition of "the nation" in Mauritius. Conventional Western descriptions of the nation commonly refer to a group of people who share a common identity, a common political ideology, language, religion, culture and history. To build a multicultural identity, traditional constructs of the nation need new foundations that include people of different origins, histories, and cultures. This approach has been the norm in multicultural societies around the world and includes many island nations such as Mauritius. In building a multicultural identity Mauritians recognize the need to carefully balance diverse parts within the whole, give credence to what distinguishes some elements and celebrate their role as part of the larger community. Cultural heritage tourism projects have the potential to raise awareness of Mauritian national identity through shared pasts and make additional progress in identifying what it means to be Mauritian.

Definitions of cultural heritage tourism have been a point of contention among scholars for many years (see Timothy and Boyd 2003; McKercher and du Cros 2002; Poria et al. 2001; Garrod and Fyall 2001; Tunbridge and Ashworth 1996). The common idea behind cultural heritage tourism is to use culture and aspects of the past as tourist commodities for the present and future. Included in this broad definition are tangible (e.g. physical structures and artifacts) and intangible elements of culture (e.g. folksongs and myths). Between national identity and cultural heritage tourism there exists an undeniable, but under-recognized and underutilized relationship (Palmer 1999). In the production of tourism, the use of historic symbols, signs, and topics form a discourse that characterizes a nation and can play an active role in nation building. In this manner cultural heritage tourism works toward the creation of a national identity that is perceived true and has relevant meaning ascribed to it.

In examining the question of multicultural identity in the case of post-colonial Hong Kong, Henderson (2001) observes that for the people of post-colonial Hong Kong, their unique identity is currently being validated through tourism. In this period of transition from British to Chinese governance, depictions of heritage through tourism point to significance and a concrete acknowledgment of an identity that reproduces Hong Kong's experience. In this situation where politics (power) and daily existence (culture) come together in tourism, meaning has been derived (Best and Kellner 1991). A seemingly satisfactory meaning has been agreed upon in Hong Kong, but power relations between those who identify aspects of heritage for use in tourism and the signified objects are ever-changing. The relationship between the signifier and the signified is asymmetric and as a result, the signified is subject to repeated re-creation and at times, contestation.

Contrary to the experience of Hong Kong, Fürsich and Robins (2004) illustrate a case where depictions of identity through tourism have not been affirmations of true identity. Focusing on how African governments use tourism websites to advance manufactured culture, renditions of history, and readings of heritage all in the name of marketing Africa as an international tourist destination, nations have introduced a discourse of heritage and identity that reaches beyond the tourism landscape into the realm of national identity. The discourse of tourism has been substituted for the discourse of national identity, running the risk that a tourism discourse which romanticizes the colonial past, furthers African stereotypes, and promotes inaccurate colonial histories may be incorporated as "knowledge" (Fürsich and Robins 2004).

According to Aumeerally (2005), the Mauritian government has taken a stance against romanticized notions of the colonial past in tourism marketing, instead opting for an image that is more inclusive of the nation's plural identities. It can be argued that this tactic has been only moderately successful since the government can only control the images it produces, leaving private enterprise the freedom to construct and disseminate marketing materials that oftentimes utilize a nostalgic approach to colonialism. Both approaches have proven to be economically viable. The government has chosen to emphasize various elements of Mauritius' diaspora depending on the target audience (Aumeerally 2005). To attract Indian tourists (or Indian investors), Mauritius' Indian diaspora is put in the spotlight, but can easily be refocused when marketing Mauritius to the Chinese. As most of Mauritius' tourists are from Europe, colonial images can be powerful influences used to exploit the European market segment. The concurrent use of images of colonial Mauritius and contemporary Mauritius helps solidify the identity of the nation at home and on a global scale by acknowledging the contributions of the past, while staying firmly planted in the present.

Forging harmonious relations between coexisting cultural groups without giving preference to one or the other is a challenging task fraught with difficulties. Mauritius has been actively constructing its multicultural identity since the mid-1960s in anticipation of independence in 1968 (Alladin 1993). The path has not always been clear or easy, but for Mauritius, it has largely been peaceful due to mutual respect between ethnic groups. Use of cultural heritage tourism, as illustrated by Henderson (2001) in the case of Hong Kong, can contribute to a positive sense of multicultural identity when the distinctiveness of a people and place are recognized by powerful actors that include government, foreign tourists, and local society. The same relationship between cultural heritage tourism and identity formation can be observed in Mauritius, but is, as Palmer (1999) suggests in other contexts, rather subtle and should be given more credence. Cultural heritage tourism has been present in Mauritius' tourist brochures since the 1950s and possibly earlier, but it has not flourished to the same extent as sun, sea, and sand tourism. The inability to arrive at a concrete definition of cultural heritage tourism, what it includes and what it does not, has negatively impacted its development and slowed legislation that designates what constitutes culture and heritage.

Developing Mauritian Cultural Heritage Tourism

In 1988, the Mauritian government convened a tourism seminar that resulted in the first White Paper on Tourism. A stated objective of the seminar was to discuss the addition of a cultural element to diversify Mauritius' tourism. Assessing the state of the existing tourism product, seminar participants representing foreign and domestic tourism stakeholders, agreed that as a group, they "[*deplored*] the absence of a real 'Culture - Industry'" (Government of Mauritius 1988, 10). It was predicted that tourists would show an increasing inclination toward cultural experiences and therefore Mauritius should encourage tourism development in the interior of the island through the establishment of guided tours. Inland tours would take some of tourism's burden away from the seashore, portray another side of Mauritian life, and hopefully redistribute some of the economic benefits of tourism. Following the findings of the seminar group the government decided to maintain the established destination image as an idyllic island resort, but to pursue "Culture Industry," meaning cultural heritage tourism avenues such as inland tourism (versus the already popular coastal tourism), building more museums and parks, and developing historic sites including the Citadel (also known as *La Citadelle* and Fort Adelaide), a British-built fort atop a hill in Port Louis. This new cultural facet to the sun, sea, and sand core product would augment destination marketing materials. Not to be taken lightly, the government vowed to "project an image of excellence in the promotion of the cultural diversity of Mauritius" (Government of Mauritius 1988, 13).

Forward movement in the area of cultural heritage tourism in Mauritius has been slow, but it is happening. Lagging somewhat behind are the regulations which would aid in the area of cultural heritage tourism development. Although legislation has existed since the Ancient Monuments Act of 1944, the protections offered have been limited to tangible elements, particularly built structures. Due to definitional constraints surrounding the concept of heritage, legislation of the past did not include intangible culture or other aspects of heritage. However, progress has been made in the form of the National Heritage Fund Act of 1997, which was replaced by a new, but similar Act in 2003. This Act is more inclusive than previous legislation and seeks "to discover the heritage which joins and not that which *divides*" (emphasis in the original document at http://www.gov.mu/portal/site/nheritage). Tangible and intangible aspects of culture and heritage are included in the 2003 Act as well as language expressing the desire to identify, preserve, protect, and develop sites.

Through legislative acts, the government is a major participant in determining what is or is not "historic" and holds the purse strings that can ultimately make the decision as to which places are funded for development or preservation. In this way, the government is legitimizing the signified. As government is commonly a reflection of the societal elite, their desires are reflected through government action. While this may appear to be a permanent one-sided determination about what is important in terms of history and heritage, there have been changes in the Mauritian elite over the years with a shift to a Hindu-Mauritian elite (Teelock 1999). In the past, the impetus for preservation and conservation in Mauritius has largely come from the Franco-Mauritian minority. This group, with their ties to Europe, high levels of education, recognized social status, and above average economic standing, has led the way in terms of establishing NGOs, conducting historical research, and filling government nominated stewardship positions with the task of identifying tangible heritage in the forms of buildings, monuments, and other structures.

Legislation and government funding alone cannot create viable cultural heritage tourism initiatives. Financial support is essential, but difficult to come by in a developing country. Informants point to a lack of capital as a reason for not developing or protecting many aspects of Mauritius' cultural heritage. Creative financial solutions are actively being sought that include government sanctioned stewardships to private entities and NGOs, foreign country support for projects, and supranational organization sponsorships.

The Mauritian public is another of the parties involved in historic preservation and national heritage designation. According to Teelock (1999), private citizens are a powerful force in support of preservation efforts and are oftentimes very much aware of the importance of preservation. In many cases, citizens are emotionally connected to place and the surrounding landscape. Involvement of the public in determining what is considered national heritage, developing the site, and maintaining it, solidifies connectedness to place and makes that place more meaningful across the board. All this translates into the messages that outsiders, domestic or foreign tourists, receive about the significance of that place (Teelock 1999).

Examples from the Mauritius Tourism Landscape

In what follows, two historic sites found in the tourism landscape of Mauritius are presented to further demonstrate the relationship between identity formation and the development of cultural heritage tourism. Each is unique in what it represents to locals and tourists about the Mauritian nation and its people. Both places are designated by the Government of Mauritius as National Heritage Sites and both have applied for inclusion on the UNESCO World Heritage Site list. One has successfully earned a place on the list.

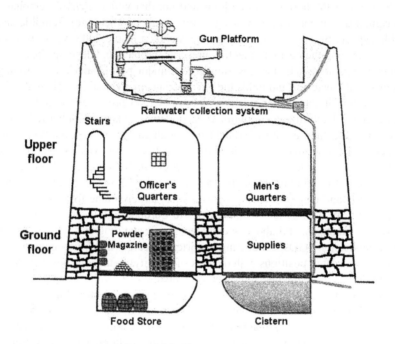

Figure 5.1 Cross-section of a Martello Tower

The Martello Tower

One way to explore Mauritian cultural heritage is by examining the preserved Martello tower, a small, conical stone fort, found on Mauritius' southwest coast. The Martello tower concept is not unique in and of itself. The original idea for the tower came from the French who in 1794 successfully fended off the British as they attempted to take the island of Corsica. It was a single tower perched north of Corsica which prevented the advances of the British, using no more than thirty men and a single cannon. The impression left was so great that the British began using these French designed towers for their own protection and this is how the towers came to Mauritius. The preserved Mauritian tower is a lesser known part of the tourist landscape as it is located in the somewhat marginalized and largely rural Black River

Figure 5.2 Martello Tower in La Preneuse, Mauritius
(by Philippe la Hausse de Laloviere)

district with a predominantly Creole population. The tower, now fashioned into a museum, is reported to be one of the best preserved Martello towers in the world (see Figure 5.1). Mauritius' preserved tower is a unique design which was conceived and built by an interesting combination of efforts. The tower contains many of the clever design elements found in other towers such as a sunken cistern for fresh water, a rain water collection method, varying wall thickness, and an effective ventilation system (see Figure 5.2). All towers were topped with a single cannon mounted to a rotating platform. The construction materials of the Mauritian tower had to be varied out of necessity. The primary material used was local basalt rock which was hand cut by Indian stone masons. Slaves were used for much of the heavy labor consistent with building a stone fort. This labor force was supplemented by British troops and was supervised by British officers.

Today the tower is still manned, only not by British soldiers, but by young adults from the local community. Uniformed guides are multi-lingual and trained in the present-day interpretation of tower history. In addition, they have been educated about the many different roles played by those who had a part in the design, construction and use of the tower. Emphasis has been placed, by its Franco-Mauritian steward, on what makes the tower Mauritian rather than British or French.

This knowledge of the tower's past, when relayed to tower visitors, allows for people from very different backgrounds and ethnic groups to see a part of their own pasts, their own heritages. The European tourist may have some preexisting knowledge of Martello towers, but even in the absence of that knowledge, can connect with the tower as an aspect of European history. Franco and Anglo Mauritians can likewise find their pasts in the tower. When the story of the tower is told in its entirety, the contributions of Blacks and Indians become apparent and a sense of belonging gained.

This approach to telling the story of the tower recognizes the importance of all who contributed to it, including the rocky island itself, and thus counters a situation which has occurred in the past when identifying objects of tangible heritage in Mauritius. An Acting Minister of Arts and Culture illustrated this point well by using the Sir Covil Deverel Bridge as an example of how historians will use the bridge to comment of the contributions of Deverel during his lifetime, but will neglect to mention the many hard working laborers, who they were or how they contributed to the building of the bridge. However well the history of the Martello tower may reflect the collective past of Mauritians, as it stands, over 45 per cent of the Mauritian public is unaware that the Black River Martello even exists. Of those who did know about the tower, only 33 per cent felt it was important to their identity as a Mauritian.

Aapravasi Ghat

Aapravasi Ghat is another of the less visited heritage sites in Mauritius. It is located on a small parcel of urban land along the main highway and across from the very busy Immigration Station bus terminal in downtown Port Louis. Although it is tucked away in a hectic part of the city surrounded by commercial interests, the site is still within easy walking distance of the festival-styled marketplace that serves as a popular shopping, entertainment and general gathering spot for locals as well as international tourists. *Aapravasi Ghat*, formerly called *Coolie Ghat*, was originally named for the unskilled Indian laborers, coolies, who landed in Mauritius' harbor (*ghats* = banks, as in river banks) in large numbers shortly after the abolition of slavery was announced. The landing station was renamed using the term "*aapravasi*" loosely translated from Hindi as "immigrant." *Aapravasi Ghat*, serving as an immigration processing area, was the immigrant's first experience of Mauritius where they were temporarily housed before being sent to plantations.

The establishment of the immigration station and the arrival of Indian immigrants corresponded directly to the labor shortage experienced as a result of the abolition of slavery in 1835 (UNESCO 2005). Keeping this in mind, the national heritage status afforded to the *Aapravasi Ghat* site means to represent all aspects of indentured servitude experienced by immigrants to Mauritius. This is not meant to lessen the experiences of Indian immigrants but to accentuate the shared experience of forced labor suffered by all immigrants. A little-known fact about the *Aapravasi Ghat* is that it was for all non-European immigrants arriving from 1849 until the 1920s (UNESCO 2005). Indian laborers made up the largest number of immigrants, but African, Chinese, Comorian and Malagasy immigrants were also processed through the depot (MGI 2 2003; Ly Thio Fane Pineo 1984).

Figure 5.3 Aapravasi Ghat Courtyard

The *Aapravasi Ghat* site had been in a state of flux for several years with no real direction for its preservation, restoration or development as a tourist venue until 2001, when the Aapravasi Ghat Trust Fund Act was passed. Over the years, the site has been protected as a national monument, and as such, various commemorative plaques and a statue have been added to the location (see Figure 5.3). As a part of a beautification project, a fountain was constructed on site and benches added and European-fashioned light fixtures installed, giving an incongruous look and feel to the site. The site had fallen into a state of disrepair which characterized the entire landing station (see Figure 5.4). In 2003, however, this changed under the guidance of the Aapravasi Ghat Trust Fund Consultative Committee. In preparation for the official visit by UNESCO World Heritage Site committee representatives, the site was spruced up and an archeological excavation took place where members of the public were welcomed as volunteers.

Figure 5.4 Stairs Entering Aapravasi Ghat (right corner) from Bay of Trou Fanfaron

The public has not shown a great deal of interest in working toward preserving the site despite the fact that nearly 70 per cent of the population considers *Aapravasi Ghat* as important or very important to their identity. The high rating of the site may reflect that people are aware of the site and its relevance to the Indian population. Judging by those in attendance at the Aapravasi Ghat Trust Consultative Committee meetings, few members of the public responded to the call for input on what should be done at the site. Other than professional middle-aged Indians, the Consultative Committee meetings have not attracted other ethnic groups or young adults. The reality of the matter is that most Mauritians feel that *Aapravasi Ghat* represents Indian heritage, not Mauritian heritage.

The task for the Aapravasi Trust Fund members is that of repackaging the immigration landing station. It will need to rewrite history and work to re-educate the public before the legitimacy of *Aapravasi Ghat* as a part of an overall Mauritian identity and the larger national heritage can be acknowledged. In the midst of that effort, there was the concurrent goal to give *Aapravasi Ghat* a more significant place in the tourist landscape by having it designated a World Heritage Site and perhaps having it included in the proposed Port Louis Heritage Trail (Mauritius Tourism Development Plan 2003). Questions still abound regarding how the site should be modified. Should *Aapravasi Ghat* be preserved or restored? How "authentic" should the site be? Achieving World Heritage Site designation in 2006 has resolved many of these questions since the site will adhere to UNESCO rules and procedures for continued development.

Conclusion

Mauritius is undergoing a change of perspective connected with the growth of the Mauritian nation. Nation-building activities are taking place in Mauritius and the development of cultural heritage tourism is one example where a more meaningful identity for the people of Mauritius is being constructed. In Mauritius' heritage tourism this is reflected through the process of supplementing, not supplanting, the well-established portrayal of heritage from a European colonial perspective and including the experiences of multiple Mauritian identities. This is a sign of societal transformation. It can be difficult, even in a place like Mauritius where multiple identities exist and are legitimized by government, to move beyond the question of whose heritage is on display and instead realize that heritage (regardless of whose) is presented for the nation as a whole to honor in the spirit of shared identity.

Pinning any degree of hope to cultural heritage tourism being the solution to nation building and refining the definition of Mauritianness would be a mistake when there is so much involved in such an endeavor. However, further development of national identity and national pride can benefit tourism and vice versa as is demonstrated by the examples of the Martello tower and *Aapravasi Ghat*. Eriksen (1997) observed circumstances where Mauritian nationalism surpasses ethnic boundaries—the continued and conscientious development of cultural heritage tourism in Mauritius can be another case where the interests of national identity and furthering Mauritianness go beyond ethnic loyalty. By coming together in solidarity, investing themselves, and taking an active role in cultural heritage tourism endeavors that embody the nation, Mauritians empower themselves and impart meaning to the tourism product.

For the relationship between cultural heritage tourism and national identity to be a mutually beneficial exchange, three steps must be taken. First, Mauritians must be educated or in some cases re-educated about aspects of their shared heritage, the specific places where events transpired, and how these places and events represent all of Mauritius. Some of this renegotiation is designed to happen in the school system, but the adult population must equally be made aware through media campaigns and other methods. Second, greater dissemination of knowledge should come from Mauritians themselves. After attending the meetings of the Aapravasi Ghat Trust Fund Committee and making site visits, it is apparent that there is a distinct lack of professionals in Mauritius who are educated and trained to work in heritage-related fields such as archeology, period restoration, and site management. In concurrence with Teelock (1999), dependence on foreign experts from the UK and India should be reduced and in cases where using outside consultants is unavoidable, efforts to encourage transfer of knowledge are crucial. Third, continued interest in producing and marketing cultural heritage tourism is essential. With the involvement of more ethnically diverse segments of the population, increased funding from government and private sources, and a broadened definition of national heritage which includes a wider variety of tangible objects in addition to intangible aspects of culture and heritage, the chances for cultural heritage tourism to be successful have increased. All levels of Mauritian society along with the various stakeholders should be encouraged to participate in the conception of cultural heritage tourism, helping to decide what aspects are to be marketed to tourists and come to some agreement concerning the accompanying messages about Mauritian culture. Mauritian enthusiasm and emotional connectedness to the tourist landscape will be apparent to both domestic and international tourists seeking to know the "true Mauritius."

Chapter 6

Slicing the *Dobish Torte*: The Three Layers of Tourism in Munich

Richard Wolfel

Introduction

One of the important debates in studies of tourism is whether space is produced for tourists based on their needs, wants and desires or whether space is constructed and consequently consumed by tourists. Those who argue tourism space is produced as a result of tourist demands suggest that tourists are the primary decision makers in the development of tourist sites. In fact, some definitions of tourism such as those by Mathieson and Wall (1982) and Ryan (1991), focus solely on the tourist and neglect to mention other participants in the tourism process. This perspective presupposes that tourists wield the most power, deciding single-handedly what sites to visit and how those sites are to be interpreted.

Such a view, of the primacy of the tourist as the consumer, can be contrasted with ideas that space can be produced by a state for consumption by a tourist. This view is informed by the critical approaches to landscape interpretation. Such views emphasize the state as the active producer of space, creating space to promote their discourse of national development. While proponents of this view, such as Peleggi (1996), Guano (2002) and Gotham (2002), generally have focused on general studies of space, connections can be made to studies of tourism. The state can use tourism as a method of transmitting its values to domestic and international tourists.

In an effort to demonstrate how space is produced for the tourist, this study utilizes Munich, Germany (München), the capital of the Bavarian region, as a case study. Germany, in general, is a useful example illustrating the production of space because the German nation has undergone several dramatic shifts during the twentieth century. Munich in particular is an exemplary case study due to its position as an important city in the re-development of German, specifically West German, national identity during the Cold War era. This West German identity is still important as it has been the dominant discourse in post-unification Germany. Munich has been viewed during the post- World War II era as one of the focal points of German development. The importance of Munich for the West German state forced West Germans to engage with issues of nationalism in the rebuilding of the city.

The problem of producing an ideal "German" tourism space in Munich stems from the history of the region. Like a *Dobish Torte*, Munich is made up of three layers of tourist themes. First, Munich is famous for the traditional Bavarian themes that are dominant in the city. Most of the *Alstadt* [Old city] was reconstructed in the post-war era to look as it did prior to 1933. Layered on top of the traditional Munich is the modern Munich, one that was designed to show the promise of the new (West) German state. The Olympic Games of 1972 were the pinnacle of such thinking.

The third layer of tourism is more difficult to view. This is the Nazi layer of Munich. Consistent with political thinking in West Germany during the Cold War, the leadership of Munich has ignored this layer. As a result, exploration of this layer requires an informed tourist who can sift through other themes to find sites that were significant in the history of the Nazi movement. Often, these sites are not prominently marked. The hidden nature of these sites is not by chance and is consistent with the geographic literature on landscape that emphasizes that the state actively creates space in an effort to promote its identity (Gotham 2002). The post-World War II identity of West Germany has never adequately dealt with the Nazi era, thus it has remained largely hidden from view.

Lefebvre and the Production of Space

A key theme within the geographic community concerns the definition of space and how it is produced. According to Lefebvre (1991, 1), until recently, space had a "strictly geometrical meaning." In other words, space was not something that was influenced by cultural or political action; it was innocent, and isolated from the political process. Under this view, space is considered to be a container, the base on which actions occur, but not part of the process. The translation of Lefebvre's *Production of Space* into English signaled a major change in the thinking about space. Lefebvre challenges the innocence of space in the political development of a region by prominently stating "that every society ... produces a space, its own space" (Lefebvre 1991, 31). No longer is space seen as simply efficient, or appropriate, but now space is viewed within a larger political discourse. Space is an active player in the political development of a region; it influences change and is changed during the process of political development. There are a myriad of influences on the production of space. Lefebvre (1991, 8) identifies a long list, including: geographical, economic, political, sociological, and ecological influences. In other words, multiple spaces are layered on top of each other. A single point in space is not a simple point. It is the result of numerous processes that work to influence the structure of society. The location and characteristics of that point are the result of many struggles between the competing factors listed above, as well as between different groups within a society each seeking to bring about their own conception of how each of these numerous processes should develop.

One of the most important societal characteristics influencing the production of space is the mode of production within a society. The mode of production can be defined as "the sum total of [the] relations of production constituting the economic structure of society – the real foundation on which rise legal and political superstructures and to which correspond definite forms of social consciousness" (Himmelweit 1991, 379-80). This indicates the economic base of society is an important influence on all aspects of society, including the design and usage of space. Therefore, any discussion of space must be situated in the society's mode of production.

By extension, it can be seen that major changes in space should occur during important societal changes. Such changes usually result from dramatic events in the history of a society. For example, Russia is undergoing one of these dramatic changes now due to the collapse of the Soviet Union. As Forest and Johnson (2002) have emphasized in their project, the architecture and monuments in Moscow are being reevaluated in this new era. Monuments can be modified to fit into the new national identity, or they can be discarded if their message is in conflict with the national discourse.

Societies produce space for very specific reasons. First, societies use space as a method of transmitting their conceptions of identity. For example, Duncan (1990, 20) contends that landscapes tell "morally charged stories." This sentiment is echoed by Atkinson and Cosgrove (1998, 32) in their study of monuments in Rome; they conclude that large, public monuments were built during the late nineteenth century in an effort to "locate and embody national and imperial identities and meanings in key metropolitan locations." Architecture can be used to identify the important elements of national identity within a place—the ideas and individuals that a nation values are commemorated in the urban built environment. The message that the architect and the leadership of a nation intend to communicate to the citizens of that nation can be "read" and interpreted by viewing these buildings and an analysis of the history of the building.

The other important use of space is to promote power. Smith (1993, 88) notes, "landscape is, in part a 'work' consisting in itself as the construction of specific individuals and parties in pursuit of specific technical, political and sometimes artistic goals." In other words, landscapes are developed in an effort to promote a certain identity, often at the expense of other potential identities. Such a conclusion leads to an evaluation of landscape that not only focuses on the message of the architect but places the individual building, or monument, in a wider discussion of the overarching political, economic, cultural, and social development of the nation. Ley and Duncan (1993, 329) emphasize this in their conclusion, "landscapes and places are constructed by knowledgeable agents who find themselves inevitably caught up in a web of circumstances – economic, social, cultural and political – usually not of their own choosing." This is an important guiding principle in modern studies of landscape within geography. Landscapes are not just isolated objects, but part of the wider process of identity construction undertaken with a specific objective in mind. Usually this identity construction is part of the process of nation-building and the urban built environment is an important component of the process of national development.

Tourism and Landscapes

One of the major issues in tourism concerns how people pick sites for touring. The homogenization of cities under globalization would suggest that people do not consider specific characteristics of cities. Chang et al. (1996) observe that the unique processes of cities are generally not discussed in tourist studies. This view has been challenged elsewhere in tourism studies. For example, Peleggi (1996) suggests that the preview of sites at home influences site selection. Richards (1996) similarly concludes that tourists are selective in their consumption of heritage resources and that tourists' selectivity has its basis in personal preferences. Indeed, sites tend to market their unique characteristics in an effort to attract tourists because increased globalization requires that sites more clearly define their locations in the global market and fill niches within the tourist industry.

Another important issue within the tourist literature concerns how sites acquire meaning. Richards (1996) suggests that tourists create their own meaning for tourist sites based on their experiences and personal characteristics. Tourism therefore is a process that involves consumption, not production.

The notion that the meaning of a tourism site is created solely by tourists has increasingly come under attack. The importance of the state as a selector of tourist sites is prominently stated by Nogue and Vicente (2004) who argue that the cultural projection of society onto a space is a fundamental element of the development of national identity (see Edwards and Llurdes i Coit 1996). Additionally, Peleggi (1996) and Guano (2002) suggest that sites are selected for tourism in an effort to promote the official historical narrative of a society and provide legitimacy for societal elites. Gotham (2002) notes that discourses on nationalism can occur overtly through the widespread promotion of an important historical site, or covertly through the use of language in brochures explaining a site. In this way, governments utilize tourist landscapes in an effort to provide a set of memories for tourists to consume and through this consumption of historical tourist sites, people are given a vision of the history of the nation (see MacCannell 1976 and Nuryanti 1996). Of course, it is important to note that if sites are used in an effort to promote one dominant discourse, other alternative discourses must clearly be omitted. Similarly, as societies develop tourist sites, some sites are preserved, others are destroyed; some sites are placed in the foreground and others are moved to the background. It is precisely this process that can be seen to occur in Munich.

The Three Layers of Tourism and the Production of Space in Post-World War II Munich

A term that has been used to describe the politics of memory that has characterized the post-World War II era in West Germany and modern Germany is *Vergangenheitsbewältigung*. This term can be translated as "coming to terms with the past." Specifically, the past that needed to be come to terms with was the contested, still evolving memory of the Third Reich (Rosenfeld 2000). This is a key idea in post-war politics in the two Germanys and the united modern German state. While the Nazi past in the former East Germany is beyond the scope of this paper (see Herf 1997), the West German process of identity construction is very important to understanding tourist sites in Munich.

This process of identity construction is especially evident in the major cities of Germany. According to Till (2005, 5), German cities are the foundation on which "German nationalism and modernity ha[ve] been staged, restaged, represented and contested." Hence, cities are the primary stage for the development of German nationalism. It is the major cities that have been important focal points through the history of the German state. During the post-war era, cities were important because of the damage they suffered and the politics involved in their reconstruction (see Ladd, 1997 for a discussion of this in Berlin).

The rebuilding of German cities can be viewed and read as a metaphor for the reconstruction of German identity. This process of rebuilding has been characterized by three rather distinct tendencies. The first of these, the "Geopolitics of Nostalgia" (Till 2005, 129), refers to the tendency to reconstruct German cities so as to look like they did prior to Hitler's ascendancy to power. A second approach has been to create post-modern spaces that make no reference to the past. In Munich, this tendency is exemplified by the *Stunde Null*, or "Zero-Point" thinking. For proponents of *Stunde Null*, the destruction of cities during World War II represented an opportunity for Germans to build new identities. Integral to this movement was the building of new, modern buildings on old sites in an effort to separate the site from the past. A final tendency has been characterized by, until recently, ambivalence about the Nazi era, an "inability to mourn" (Mitscherlich and Mitscherlich 1975). This inability to mourn has resulted in Nazi era sites often remaining unmarked as if to ignore or disavow the Nazi era (Crump 1997). Each of these tendencies characterizes a distinct layer of tourism in Munich and each is discussed in more detail below.

Layer 1: Traditional Munich and the Geopolitics of Nostalgia

Layer one represents the traditional landscapes of Munich. These landscapes were strongly influenced by a geopolitical climate of nostalgia that sought to ignore the Nazi era by rebuilding Munich to look as it did prior to the arrival of the Nazi regime. What makes Munich the most frequented tourist destination in Germany is this traditional layer of tourism. This layer portrays the glory of Munich as the former capital for the Wittlesbach dynasty, the family that ruled Munich until its incorporation into the German empire, and it is centered on the preserved and reconstructed *Alstadt* of Munich. While the preservation and reconstruction of the *Alstadt* has been referred to as an example of historic preservation, Rosenfeld (2000, 29) suggests that it may also be seen as a method of "revising history and forgetting the past."

The process of rebuilding included exact reconstructions of important sites to look as they did before the war. Rosenfeld (2000, 32) identifies several of these, including the *Peterskirche, the Frauenkirche, the Rauthhaus, Glockenspiel*, the *Residenz*, and the *Hofbraühaus*. These buildings were significant sites in pre-war Munich. The *Peterskirche* was the oldest church in Munich, the *Frauenkirche* was the major cathedral of Munich and the *Residenz* was the residence of the Willesbach family.

The *Hofbraühaus* (see Figure 6.1) is one of the most famous symbols of Munich. Built in 1589 by Duke William V, this is the classic beerhall that has made Munich an important tourist destination. This large building is an important and highly publicized destination for tourists in Munich looking for an "authentic" Bavarian experience. The *Hofbräuhaus* features a traditional Bavarian polka band, waitresses dressed in traditional Bavarian *dirndls* selling large pretzels and mugs of beer, and locals dressed in *lederhosen* drinking from beer *steins*. Traditional Bavarian food is served on long tables, where complete strangers are seated next to each other.

Figure 6.1 The Hofbraühaus in Munich

The most prominent celebration in Munich is the *Oktoberfest*. This tradition, originally held to celebrate the wedding of a Bavarian prince, has become a tourist event of global prominence. The celebration is held in a large urban park, maintained specifically for the *Oktoberfest*. The major breweries of Munich construct purposely built pavilions and provide entertainment, food, and beer. This is an opportunity for Munich to show itself as a traditional European city, all while distancing itself from any connections to the Nazi era.

Layer 2: Modern Munich

The second layer of tourist landscapes in Munich stands in direct contrast with the landscapes of the geopolitics of nostalgia and represents landscapes strongly influenced by the *Stunde Null* movement, which believed the destruction of German cities during World War II represented an opportunity for Germans to build a new identity. This second layer characterizes development on the outer edges of Munich, away from the traditional center. The new Munich is associated with the West German economic "miracle" to which Munich was an important contributor. The post-war era saw Munich in particular, and Bavaria in general, emerge as the high technology center of Germany. Major corporations, like Siemens, Audi, BMW, and Allianz, built their headquarters in Bavaria. This new era demanded new architecture, which then became an important part of the visual landscape for tourists visiting *Munich*.

The show piece of this new thinking was the stadium constructed in anticipation of the 1972 Summer Olympics in Munich. The *Olympiazentrum* was built to show the new Germany to the world. The site is serene, with a strong tie to nature. This stands in contrast to the militaristic, nationalistic landscapes of the 1936 Berlin Olympics. The Munich *Olympiazentrum* was built to demonstrate the themes of Adenauer's "*Westintigration.*" According to this belief, West Germany was a new state, one that acknowledged the horrors of the past, but saw the new democracy as a break from these negative historical events. The emphasis on a new era, dominated by a new type of thinking, democracy, would allow for the citizens of West Germany to "forget" about the past. The major theme in the architecture of the *Olympiazentrum* is the alpine landscape that stands to the south of Munich. Figure 6.2 shows the *Olympiazentrum*, with the rising and falling roof structure that mimics the peaks of the Alps. This tie to nature and modern thinking was important to show Germany as a modern nation, one that has left behind the barbarism of previous eras.

Figure 6.2 The Olympiazentrum

Layer 3: Munich

The third layer of tourism in Munich represents an attempt to engage with the Nazi era. In the reconstruction of several German cities, there has been the utilization of historic sites to admit guilt (Till 2005). Till emphasizes this as an important theme in Berlin, but such a theme is not seen extensively in Munich. Munich is one of the few major cities in Germany that did not preserve any war ruins. Given the centrality of Munich to the Nazi movement, dealing with the Nazi landscape was key to the reconstruction of Munich. There were two basic choices for Nazi sites: destruction or de-Nazification. De-Nazification involved the removal of symbols and a re-orientation of the history of a site to a prewar identity. At sites where the Nazi identity was especially troublesome, the site was destroyed.

The treatment of Nazi era sites in Munich, once termed the Capital of the Movement (*Haptstadt der Bewegung*) by Hitler, is significant. The *Führerhaus* is one example of a building that was de-Nazified. Today, part of the School of Theatre and Dance, Hitler's Munich office has no identifying placards. Since the building survived the war relatively intact, the building was stripped of its Nazi symbols and turned over to the city for use. While the display of Nazi symbols is illegal in Germany, and thus the stripping of symbols from the building amounts to usual post-war practice, the absence of a placard of any kind is a significant omission.

Figure 6.3 Bayerische Gemeindebank (Former Site of the Wittlesbacher Palais)

The *Wittlesbacher Palais* (see Figure 6.3) is an example of a site that was destroyed during the war and not rebuilt. The building was one of the palaces for the Wittlesbach family, the ruling family in Munich. During the Nazi era, the *Wittlesbacher Palais* was commandeered by the Gestapo for its Munich headquarters. With the building's destruction during the war, a decision needed to be made regarding the future of the site. Given its connection to the Nazi era, the building was not rebuilt but it was turned over to the *Bayerische Gemeindebank* for its Munich headquarters. The only clue of the site's history is a plaque on the wall (which mentions the building's significance during the Nazi era) and a replica of a lion statue that was one of the trademarks of the *Wittlesbacher Palais*.

Other portions of the Nazi landscape remain unmarked. *Sterneckerbräu*, the site of the first meeting of the Nazi party attended by Hitler, is no longer a beerhall and the history of the site is not marked. The *Bürgerbräukeller*, the starting point of the Beerhall *Putsch* is now a hotel, with no marking as to its sinister past.

The city has made some progress on memorializing certain events during the Nazi era, though these have often been the least controversial events. In the *Hofgraten*, a monument to the White Rose movement, an anti-Nazi group, was constructed. At the *Olympiazentrum*, a monument to the civilian bombing victims was constructed on a rubble hill. Elsewhere in Munich, new monuments focus on soldiers who died during the war and the victims of the Nazi movement in *Platz der opfer des Nationalsozialismus* [Square for the Victims of National Socialism]. Recently, an eternal flame was added to this square, across from the site of the former *Wittlesbacher Palais*. While not as visible as memorials in Berlin, this represents the first attempt at engaging with and acknowledging Nazi atrocities in the city.

For the casual tourist to Munich, the Nazi landscape is generally extremely difficult to gaze upon since almost all buildings that survived the war provide no recognition of their Nazi pasts. Several key historical sites have been rebuilt to disavow the Nazi past, either through a lack of plaques or through new construction. This has made tourism focusing on the Nazi era a research-intensive process. In order to find all of the key sites for Nazi history in Munich, the tourist would need to conduct an extensive amount of pre-travel research.

Conclusion

Munich provides an example of how tourism can be used to promote a discourse on national identity. Post-World War II West Germany was faced with the challenge of resurrecting the German nation from the ruins of the Nazi regime. The reconstruction of sites and identity in West Germany must be situated within the larger context of placing the two Germanys in the conflict of the Cold War. As the frontline of the East/West conflict, West Germany was in a precarious position militarily, socially, politically, and economically. In order for the fledgling West German state to survive, it needed Western support.

To gain this support, Adenauer needed to show the world that West Germany was a new, democratic state, that has evolved from the nationalistic, militaristic history of the German Empire and Nazi era. Given the volatile nature of a discussion of

National Socialism in general and the Holocaust in particular, West Germany followed two important trajectories. Each of these influenced the urban built environment of Munich and provided important tourist sites within the city.

The geopolitics of nostalgia has influenced a vast amount of construction in the A*lstadt* of Munich. This discourse on German nationalism seeks to disavow the Nazi era by rebuilding the city to look as it did prior to the Nazi era and therefore not acknowledging the memories associated with certain sites. Within the *Alstadt*, the reconstruction of the *Frauenkirche, Peterskirche, Rathhaus* and *Hofbräuhaus* represent examples of the primacy of the geopolitics of nostalgia in the inner city of Munich.

This is contrasted with the *Stunde Null* proponents who believe that the destruction of World War II presents the German people with the opportunity to rebuild their nation into a democratic, modern nation. Their intent is to build anew, thereby disavowing the past. The outer regions of Munich became the economic center of Bavaria and one of the growth poles for the German "economic miracle." This region is characterized by modern construction that is heavily influenced by *Stunde Null* thinking and is best represented by the construction of corporate headquarters, like BMW or the *Olympiazentrum*.

Both of these discourses ignore the Nazi history of Munich. The Nazi sites have either been stripped of their Nazi symbols and remain anonymous, rebuilt to look as they did prior to the Nazi era, or the most sinister sites have been rebuilt in a completely different manner in an effort to remove memory from the site. Examples of this new construction include the construction of the *Bayerische Gemeindebank* on the site of the *Wittlesbacher Palais* and the construction of a hotel on the site of the former *Burgerbräukeller*. For the tourist to locate the Nazi landscape and understand its meaning considerable research must be undertaken prior to arrival in Munich.

Traditionally, studies of tourism have focused on understanding how the tourist selects, understands, and consumes sites. Very little work has attempted to connect the discourse of national development to studies of tourism. According to Lefebvre (1991), all space, including tourist space, is produced for the societal elites. It follows that these elites use tourism as another tool in developing discourses of nationalism. Just as history and language can be manipulated to promote a particular vision of nationalism, certain sites, events and monuments are selected as significant and worthy of transmission to increase the legitimacy of the ruling regime. Tourism is an integral part of any nation's discourse on nationalism. Germany provides an example of how sites can be manipulated to promote the legitimacy of a newly constituted West German state and to distance this new state from horrors of the Nazi regime.

Chapter 7

A Nostalgia for Terror

Michelle M. Metro-Roland[1]

Introduction

This chapter is about space—the physical space of location, the manipulation of interior space to transmit information, and the distance that separates meaning from understanding. The objects for this meditation on space are two museums located in Budapest, Hungary; two museums which attempt to organize the collective memory of real existing socialism. While both offer a critique of communism, their approaches are as different as their locations within the city and their positions along the ideological spectrum of contemporary Hungarian politics.

The Statue Park is an open air museum on the outskirts of the city, supported by the left-of-center mayor of Budapest. It is a repository for a small number of the thousands of statues, plaques, busts and sculptures that were utilized by the communist party to inscribe meaning into space. The Park was opened in 1993 on the second anniversary of *Búcsú*, the final withdrawal of Soviet troops from Hungarian soil.

The House of Terror opened its doors in 2002 with the support of then Prime Minster Victor Orbán, the leader of the right-of-center *Fidesz* party. The museum is located at 60 Andrássy Avenue, on one of the grandest boulevards in the inner city where, in the 1930s and 1940s, the fascist Arrow Cross Party set up its headquarters and used its cellar as a political prison and torture chamber. When Hungary was "liberated" by Soviet troops, Hungarian communists took over the building and it became the headquarters for the State Security Authority, the *ÁVH*, who used its cellars in much the same way as their fascist counterparts, though they had need for more room and connected the underground spaces along the entire length of the block. According to the wall plaque in the museum, after the failed 1956 revolution the *ÁVH* moved out, the building reverted to a more benign existence and the cellars were eventually rehabilitated for use by *KISZ*, a communist youth organization.

Both of these museums depend heavily on the organization of space to tell their tales. It is not an overstatement to say that the meaning of each museum is directly related to the manipulation of space within each. In the Statue Park the organization of space *is* the meaning. In the House of Terror the crafting of space is a critical tool

1   I would like to thank the US Department of Education for a Fulbright-Hayes Grant which enabled this research to be undertaken. I would also like to thank Mária Schmidt, the director of the House of Terror for generosity in speaking to me and allowing photographs to be taken of the museum, and Ben Frieday at Absolute Walking Tours in Budapest.

which helps push the narrative to its teleological conclusion, the laying of blame upon a select few in the "Hall of Perpetrators" and the eventual coming up along the stairs into the clear light of contemporary post-communist Hungary. Equally important is the actual location of each museum within the urban fabric; the Statue Park being relegated to an outlying district and the House of Terror occupying a prime spot along the major cultural axis of the inner city. These locations, periphery and center directly affect their ability to influence how the story of communism will be remembered and how it will be transmitted to outsiders.

Space does not simply operate as a neutral container. As Lefebvre (1991) has potently argued, space is a social production and the intersection of space and ideology is not accidental. However, a distinction must be drawn between the official space created and maintained by state power, what Lefebvre refers to as "abstract space," and the lived, quotidian experience of space made by people known as "social space." The manipulation of space by communist regimes in the Soviet Bloc was a major plank in the creation of "workers' paradises." These included the building of efficient mass transit systems for the people, the appropriation of aristocratic properties, the construction of multi-story panel flats, houses of culture, and large-scale monuments to the workers and the heroes of the workers' movements (Crowley and Reid 2002).

For post-socialist governments, the problem became one of reuse of form and reshaping of content within the socialist cities these governments inherited. While panel flats and subway systems retained their use value as is, houses of culture could be given a slight makeover and some confiscated properties could be restored to the descendents of their pre-war owners. A larger, though in some ways easier problem was posed by street names, statues, plaques, and monuments; larger in that often the size of these monumental objects was massive, simpler in that they could merely be removed from sight/site in the case of physical artifacts and in the case of street names, the scale of change was large but the implementation required only new signs, new maps and for a time, a key to translate between the old and new (Foote et al. 2000). The rush to embrace multiparty democracy, market economies, and EU membership, and alterations to urban space marked these countries as post-socialist. Nevertheless the recent communist history still lingered like a specter haunting the collective memories of these states and the question as to how the communist past should be remembered came quickly to the fore. In Hungary, the answer to this question was quickly taken up by the establishment of the Statue Park and later by the House of Terror. These two solutions represent the vast space between the left and the right which has opened up within post-communist Hungary.

Locating

Location, location, location. The House of Terror has it, the Statue Park does not (see Map 7.1). Consider how the locations of each within the city are described.

The bilingual English-Hungarian brochure for the House of Terror says:

> Whoever has visited Budapest before knows that one of the most beautiful thoroughfares in the capital is Andrássy Avenue. The tree-lined street, with its lavish residences and

stately apartment buildings, connects downtown Budapest to Heroes' Square. It was named after one of the Austro-Hungarian Empire's greatest Hungarian statesmen, Count Gyula Andrássy. Interestingly, the 20th Century terror regimes, the Nazis[2] and Communists, both decided on a residence located on this boulevard as the headquarters of their executioners. [Terror Háza n.d.]

While the English language catalogue for the Statue Park describes the site thus:

In the outskirts of the Hungarian capital, Budapest-although not very far from the city center-can be found Statue Park, this typically Central European, yet universally unique collection of former public statues, which used to be stationed in the city's public domains, in accordance with the guidelines and the requests of the Socialist culture-politics and ideological system. [Statue Park n.d., a]

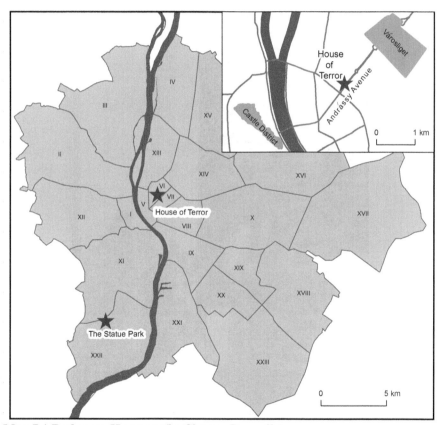

Map 7.1 Budapest, Hungary (by Shanon Donnelly)

2   In the Hungarian text it is more precisely rendered as *Nyilas*, the Arrow Cross, which was the Hungarian fascist organization, equivalent to, but different than the National Socialist party (Nazis) of Germany.

Locating the House of Terror

The opening of the House of Terror sought to reassert the presence of this former villa into the urban fabric, to juxtapose the banal with the barbarous (see Figure 7.1). Black metal cornices now decorate the building, jutting from the roof parallel to the sidewalk, giving the impression of an oversized children's stencil. When the sun shines the shadows from cut outs of the Arrow Cross and Communist Star along with the word "Terror"[3] darken the building and the sidewalk, depending upon the time of day. Another black panel extends from the roof, perpendicular to the building. It comes down to meet the ground creating a type of doorway through which the pedestrian must pass. The building has become "a sculpture of terror, a memorial to the victims" (Terror Háza n.d.).

Figure 7.1 House of Terror, 60 Andrássy Avenue

---

3   The word "terror" is spelled the same in English and Hungarian.

As noted above, the villa is situated along the main artery of the Pest side of the city. This area has been incorporated into what is termed "The Cultural Avenue" (see www.sugarut.com), a path that stretches from the castle district on the Buda side, across the Chain Bridge and along Andrássy Avenue, a recently declared UNESCO World Heritage Site. The concept of the Cultural Avenue was put forth as a way to organize the visitor experience of Budapest in a more coherent way, offering a comprehensive sampling of the variations in Hungarian culture, history, and tradition as represented by the sites located along the route (Puczkó and Rátz 2006).

To put the House of Terror into context, along Andrássy or just off it are seven museums, five theatres including the State Opera House, and two historic cafés. Andrássy ends at Heroes' Square which houses the millennium monument, two more museums, and serves as the gateway to the *Városliget* which itself contains two museums, a lake, the zoo with its listed National-Romantic and Art Nouveau buildings, and the *Széchenyi* thermal bath complex.

The House of Terror is thus positioned within the cultural heart of the city with easy access for both locals and tourists. Its emergence is part of the transformation of urban space that came with the collapse of the communist regime. The reclaiming of space took two forms. There was a revealing of the once secret and oppressive landscape as in the case of the House of Terror and the numerous memorials to the victims who fought on the side of the revolution in 1956 (see Dent 2006 for a guide to the sites of 1956). On the other hand there was a repression, an erasure from the civic landscape of the memory of communism and its heroes through the removal of offensive statues and plaques and the mass renaming of streets and squares (Gábor 2002).

Locating the Statue Park

It is because of this erasure of monuments that the Statue Park was designed. The idea first arose in an essay published during the twilight of the communist regime, but it took on a new urgency with the actual fall of the government when the future of these monuments was in jeopardy. The controversial nature of creating a park dedicated to these symbols meant that finding a location within the city was difficult. Each of the 23 districts within the city functions fairly independently of the mayor's office and though a number of possible places were eventually offered, it was the spot in District XXII that was chosen. And despite the claims from the guide to the museum seen above, the Statue Park is in fact very far from the cultural heart of the capital.

A giant backlit billboard on the way down to the subway platforms at *Deák Tér*, the meeting point of the three metro lines, and a paper poster on the platform of the Yellow Millennium Line which runs along the length of Andrássy Avenue are attempts to insert the park's existence into the inner city.[4] The billboard is positioned above the first set of escalators leading down to all the lines and shows one of the

---

4 Interestingly, those responsible for the House of Terror in summer 2006 opened another museum dedicated to communism in the town of Hódmezővásárhely, a (Hungarian) billboard for which sits on the highway just outside the Stature Park entrance.

most iconic statues from the park set against the sky at sunset. The text is first in English and then in Hungarian and the meanings vary slightly but significantly. The Hungarian says *Ma már történelem*, "today already history" while the English says "This is the history." The paper advertisement is only in English and advertises the Statue Park and the nearby stop for a direct bus line to the museum. The choice of language in the advertisements is not accidental as the park appeals primarily to foreign visitors.

The Statue Park is located in a residential and light industrial area; the parking lot is unpaved and signs leading to it are sparse (see Figure 7.2). Directions via public transit are convoluted and the round trip takes almost more time than it takes to go through the park itself. It can be a daunting experience for visitors not fluent in Hungarian. A new express bus service from the city center has been started though it goes only two times per day, and the inner city stop is not obviously marked.

Figure 7.2 Gate of Statue Park

The park can also be visited on a number of guided tours offered to tourists including Cityrama and Absolute Walking Tours' "Hammer and Sickle Tour," which promises to show visitors communism "the way it was, comrade." The trip to the Statue Park is combined with a visit to the tour company's model communist flat where one can "touch and feel the real communist décor" and partake of Hungarian "spirits" (Absolute Walking Tours, n.d.).

The Statue Park occupies a strange position within Budapest society. Hungarians, even those who consider themselves to be on the left, are rare attendees, and are often bemused when they discover a foreigner's desire to visit the park.[5] As indicated by the predominance of English in flyers and billboards advertising the park, most visitors to the park are tourists like those who want to experience communist kitsch

---

5   Light (2000) recounts similar reactions from Hungarians.

on the Hammer and Sickle Tour. As Ákos Eleőd, the architect of the site foresaw "Inevitably the American tourist who has only read about 'dictatorship' will respond differently from the one who was bound to the region by tragic fate and who brings to the park the drama of a whole life wrecked in the name of whatever these statues stood for. Silence is what can be shared" (Boros 2002, 6).

Looking

It is not just the physical locations of the Museum and Park which stand in opposition; the entities themselves, their approaches to truth and ideology, and their use of space as a narrative tool also engender a set of binary oppositions. The House of Terror is set in a multi-storied villa, with bounded corridors which direct the narrative in a teleological manner; it offers a high-technology multi-media presentation, and is richly endowed. The Statue Park is a flat, open air museum with multiple open pathways, it offers a low technology display of simple statues and plaques, and is run down and poorly funded.

Looking at the House of Terror

Although Andrássy Avenue is one of the grandest boulevards in Budapest, it is also a major artery for traffic going out of the city to the M3 highway and is thus a busy thoroughfare. So when one enters the House of Terror past the uniformed guard and through the double doors, it is with a sense of entering a sanctuary. Somber music plays and at the end of the entryway are two stone panels, one in black with the Arrow Cross symbol, the other in red with the Communist Star, each inscribed in Hungarian respectively, "to the memory of the victims of the Arrow Cross Terror 2002," "to the memory of the victims of the Communist Terror 2002." In front of these, candles burn and a wreath, wrapped with ribbons in the tricolor of the Hungarian flag, hangs between them. In the entrance hall to the left, the air is filled with the sound emanating from a flat screen television; a man, in tears, cries out in Hungarian asking why it had to happen while his lament is subtitled in English. The ticket counter lies ahead, the cloakroom and café to the left side, and to the right, through a series of doorways, one can see a full size Russian tank sitting in a black reflecting pool while the elevator shaft, which runs the length of this four story interior atrium, reflects back the black and silver images of the *Áldozatok*, the victims (see Figure 7.3).

The story begins in the room on the third floor called "Double Occupation," which sets the stage for the subjugation of Hungary by the Nazis and the Soviets. Black and white pictures and film clips depict images of Hitler, Bergen-Belsen, and the physical destruction of Budapest by the Red Army. Telephones along the wall offer sound bites from the period including songs, political speeches from the Hungarian Regent Miklós Horthy and offerings from radio in Hungary and Moscow. Actually listening to these is made somewhat difficult by the loud techno style music which blares in the room. This piece, designed specifically for this display, combined with the shifting map of Europe displayed on four flat screen televisions on one wall and the moving images and belligerent speeches set in the middle panel which play when the music stops, induce a sense of disorientation. In many ways this room is critical for understanding the overall message of the museum as it sets up the framework for viewing Hungarians in general as the victims of outside forces, the Nazis and the Communists, in spite of the role of Hungarian fascists and communists. Without the occupying forces of the *Weirmacht* and the Red Army, the Arrow Cross and the Communists would not have been able to come to power.[6]

Figure 7.3 House of Terror Atrium

---

6   The information texts which accompany the rooms make clear the argument being made in the displays that it was foreign presence and foreign influence which made these two periods possible. The text from the "Hungarian Nazis" in the "Hall of the Arrow Cross" states explicitly that in spite of the popularity of the anti-Semitic Szálasi and his Arrow Cross party, "without German help and support, however, it could never have become a potential governing factor" (Schmidt 2003, 10). The room "The Fifties," for example, over and over posits the presence of the Red Army, the interference of the Allied Control Commission, and direct Soviet involvement as the reason for the eventual implementation of Communist Rule. These include the demand that the Hungarian Communist Party, which received just under 17 per cent of the vote hold the Interior Ministry, and the abduction by the Soviets of the head of the Small Holders Party, the main opposition (Schmidt, 2003, 16). The presence and involvement of the Soviets is also made manifest in "The Room of the Soviet Advisors" (Schmidt 2003, 21-2).

The organization and design of the museum was under the control of the Director, Mária Schmidt, working with a large team of people from musicians to architects. Because they were working in the context of an existing site, she wanted to make as few changes as possible in order to show how an ordinary place where ordinary things could happen could also be used by a totalitarian regime as a house of terror (Schmidt 2006). What has resulted is a creative use of existing space along with spatial metaphors to propel the story forward. One of the critiques has been the focus on the aesthetics rather than the facts and one can see this in the use of dramatic lighting, sound (both newly commissioned and contemporary to the period), and the overall manipulation of the entire space of an area to set the scene.

Figure 7.4 The Gulag, The House of Terror

As one enters the "Gulag" room, for example, the visitor is confronted with one of the most intense uses of spatial imagery (see Figure 7.4). The floor is carpeted with a map of Eastern Europe and the Soviet Union and the walls are covered with wooden slats simulating the interior of a cattle car. At each of the narrow ends are photos depicting the Soviet labor camps. Across the floor map, conical vitrines rise up from the ground at the sites of the camps, displaying the tokens retained by some of those interned; letters, remnants of clothing, hats, and metal cups.[7] What makes

---

7   Replicas of these metal cups with the House of Terror logo are available for purchase in the gift shop.

the room even more powerful are the flat screens, four on each of the long walls. Depending on the moment, the tortured account of a former prisoner or the grief stricken remembrance of a mother whose son was sent to the gulag, fill the room. While one face speaks the other screens remain silent, and then the scene changes suddenly. The screens shift to a barren snow covered landscape, the rattle of a train rolling along the tracks fills the room and the visitor is moving eastward toward the gulag. And then just as quickly the scene changes and the authentic victims are returned to narrate their histories in their mother tongue. Though the stories remain untranslated for the tourist, the emotions and grief still come through.

The duplicity of space under the communist regime is dealt with explicitly in several of the rooms. As one enters "The Fifties," the brightness stands out compared to what had come before. The walls are painted red and the lighting is intense. A series of booths allow visitors to sit down and watch news clips and Communist propaganda films. At each place small blue papers, replicas of the ballots utilized in the 1947 election (Romsics 1999), are available to take away or perhaps to present to the ballot box which sits in the middle of the room, before a red dais upon which hang paintings of Stalin and Lenin with Rákosi, the Hungarian leader in the center. The paintings, the dais, and the chairs upon them are all distorted in their aspect ratio, stretched lengthwise, and shrunk widthwise, an indication that all was not as it seemed. Although large scale paintings showing happy socialist realist scenes cover the walls, hidden behind the dais hang military outfits, earphones, and listening and recording devices. So, while all was well for party members, "military discipline ruled in the one-party state, and it progressively extended over the whole society" (Schmidt 2003, 18).[8] Such was the duplicitous nature of the communist state, according to the House of Terror.

This theme shows up many more times in the displays of the museum. In "Gábor Péter's Room," the head of the political police in the 1950s, the office which features a desk and chairs, and unfurnished cot, is split in two. One side is painted a calming cream, a type of wainscoting and garland ceiling moldings skirt the room, while the other side appears in harsh metal tone, the ceiling stripped away to show the pipes and wiring which sit behind the façade. Small television screens set into both walls narrate the trials and imprisonments of former party members, purged and often rehabilitated as the winds of the party leadership, taking clues from Moscow, blew. Additionally, in the room "Travesty of Justice" dealing with show trials, a "hidden" niche in the book shelf holding dossiers provides a glimpse into "the practice of arbitrary justice" administered in the trials of the 1940s and 1950s where the results were determined behind the scenes and often prior to the trial itself (Schmidt 2003, 41).

Because many of the events of terror which are narrated in the museum either happened there or had their origins in the headquarters of the political police, the museum itself is conceived as a memorial. That the House of Terror can be both a memorial and museum is possible, according to the director, because of the site itself (Schmidt 2006). This connection is made in the spaces of the museum dedicated to torture. Bars allow the visitor to see into the stark cubicle entitled "Torture Chamber,"

---

8   Ironically, in the corner of the room, hovering over the surveillance equipment from the *ancien régime*, stands a security camera utilized by the current occupants of 60 Andrássy.

which sits next to the modern restrooms. The visitor is told that "this room was preserved in its original form. At the time it was referred to as the gym" (Schmidt 2003, 30). The fact of authenticity, of continuity, is at first striking and is perhaps only marred by the addition of "instruments of torture" which have been provided to enhance the experience for the museum visitor of this rather banal space containing a faucet and floor drain. Hungarians and those who purchase the audio commentary are treated to a narrative of one man's experience of torture. It is a story difficult to listen to and adds yet another patina of authority to the visual.

Perhaps it is too much of a critique that the torture victim does not make specific mention of that particular room but the lack of mention does raise the issue of how effectively places can speak for themselves. It is obvious that the drama and effect is heightened by allowing this story to be told here. But this fact simply begs the question of what role a historical museum should play—one of archiving things or places, or crafting things and places into a dramatic story with beginning, middle, and end, utilizing even slightly anachronistic or atopic, out of place, tools to that end. The materials in the museum range from historical artifacts to "representative examples," with no actual historical value (Schmidt 2006). And even more often the things which fill the rooms are contemporary artifacts made as props to move the narrative along.

This tension of how to make manifest what has been hidden, is perhaps most felt in the "Reconstructed Subterranean Prison." The basement had been reconverted over time so that in the 1990s, when the building was acquired for the museum, there were no signs of what had been there. Yet the existence of torture chambers and prison cells was a crucial aspect in understanding how "a normal building in a normal part of the city" could be turned into a House of Terror. There were no drawings or plans from this earlier period but there were the memories of people still living who had spent time there and it was from these that the cells were recreated (Schmidt 2006). Of course what can be made is only a composite based on numerous accounts. Here the visitor can enter the cells, can feel the dampness and can make an easier leap of the imagination when the props are supplied. But it may be said that spaces such as the "Treatment Room" which displays "types of instruments of torture" (Schmidt 2003, 56) are overdetermined.

One of the most manipulative uses of space is the path which brings the visitor to the "Reconstructed Subterranean Prison." The first floor displays come to a sort of dead end, the only way out is to retrace one's steps or get onto the elevator which will take you to the final floor of the exhibit.[9] As the elevator begins its achingly slow descent to the basement a man appears on a large flat screen television and matter-of-factly describes his task of cleaning up after tortures. This is a stunning set up, further confusing the lines between preservation of fact, transmission of information and dramatic narrative. When asked about the feeling of claustrophobia that is evoked in this space and others in the museum, the director responded that everything that the visitor feels, including discomfort, is intentional (Schmidt 2006).

---

9 There is a stairway but it is unmarked and access is only granted in extreme circumstances (such as a parent looking for a lost child) by the uniformed museum employee who ushers people into the elevator.

In spite of the fact that the museum aims to make manifest the atrocities of the communist period, it also offers a striking study in aestheticizing terror. Brightly colored posters from the 1950s fill one room advertising *Bambi* soda and *Állami Áruház* (literally the State Department Store); next to it all in silver and grey, Hungarian bauxite and the "drab, shoddy utensils ... characteristic elements of the period's daily life, the determinants of the general mood" (Schmidt 2003, 46) are displayed bathed in a cool blue light accompanied by the rhythmic sound of a pick axe on stones. The blue light fades and the pile of stones in the middle of the room grow a rich orange-brown color.

Further on, the beauty in the room on "religion" is in dissonance with the message of persecution. The ceiling curves down to meet the floor; niches set in the walls display the artifacts, prayer cards, rosary beads, stoles, of the clergy. In the center, the parquet floor is stripped away revealing an enormous glowing white cross embedded in the ground and running the length of the room. At the far end an elaborately embroidered vestment hangs in a vitrine, and behind it the "large grey loudspeakers ... recall the period's blaring propaganda" (Schmidt 2003, 48). The shape of the loudspeakers artfully mimic that of church bells.

And in the end, past the torture chambers, the "Hall of Tears" lists the names of those executed for political purposes between 1945 and 1967. The names are cut out on a thin strip of black metal that is back lit and runs the extent of the room. Supporting columns vie for space in the darkened room with spindly crosses (though on some of these the Star of David is lit, rather than a simple circle). It is macabre, dramatic, and beautiful.

As one leaves, the story is told but for one small detail—the blame assigned to those who perpetrated these crimes upon the Hungarian people. The "Perpetrators' Gallery" is small, little more than a hallway, not an afterthought as much as an attempt to keep the focus on the victims. "The story we are telling is the story of the victims but we also try and show the perpetrators. There is no criminal act without perpetrators" (Schmidt 2006). However, it may be among the more controversial spaces of the museum, one of the most critiqued due to the inclusion of people still living. Because there was little reckoning after the political changes and few consequences for former party members, other than the fact that they often were best positioned to reap the rewards of privatization, the House of Terror and this wall are attempts to call to account those of the previous regime. This is a result of and a consequence of the support for the museum by the right wing *Fidesz* party which is the main opposition to *MSZP*, the socialist party and successor to the former communist party (K. Horváth 2005).[10] It is thus no accident that the House of Terror is a polarizing force in Hungarian cultural circles for the way in which it has chosen to remember the communist period, or at least the days before Hungary became the "happiest barracks in the communist bloc."

---

10 The current embattled Prime Minister from *MSZP*, Ferenc Gyurcsány, came to power in the party as a result of a calling to account of the previous Prime Minister Péter Medgyessy for his role as a state informant under the communists. Though it has also cut the other way as well; prior to that Zoltán Pokorni, a member of the right of center *Fidesz* resigned his party position when it was revealed that his father had worked for the secret police.

With any historical event, the remembering and the retelling will afford many, and often divergent stories. If we go as far as Hayden White, all historical narrative is a story and historians follow the same framework for telling that narrative as any storyteller, offering up romance, satire, comedy, or tragedy (White 1973). While the House of Terror clearly sets up the story as a romance, the triumph of good over evil, the Statue Park could be said to offer up a satire. In many ways it stands in stark opposition to the House of Terror and yet it too offers up a critique of the Communist era, though in a less teleological and doctrinaire manner, offering a subtle critique which is missed by many.

Looking at the Statue Park

The Statue Park sits uncomfortably in an unpaved lot (see Figure 7.2). Set in niches on either side of the main brick gate are block style stone statues depicting Lenin (#1) and Marx and Engels (#2).[11] A small gate to the left side of the main gate leads to the ticket counter. The glass box counter is festooned with souvenirs available for purchase; compact discs of "The Best of Communism," wax candle busts of Stalin, Lenin, and Rákosi, postcards of the park and well known communist propaganda posters. On the opposite wall hang a large number of t-shirts with ironic references to communists and communism. Also available for purchase is the guide to the museum in English, French, German, Italian, and Hungarian. The entrance fee is quite cheap, 600 forint, and the catalogue is another 600.[12] The catalogue is critical to gaining a full sense of the subtleties of the museum design but many forego the guide or read only the bits about the statues rather than the full text laying out the rational behind the park design.

The entrance to the interior is through a high-walled entryway reminiscent of the entrance to a stadium. Inside is a wide open expanse comprising gravel paths, grassy beds, and brick plinths and walls in three sections. There are no directional signs and no obvious way to proceed into the displays but most visitors begin with one of the two circular paths that branch to the left and right of the main path (see Figure 7.5).

The park's collection consists of statues, from individual busts to large monuments. There are also a number of plaques which were ubiquitous features on Budapest's buildings. Plaques in the Statue Park recognize the residences of heroes of the communist movement during the interwar period (#28), the site of the first district's workers' and soldiers' council from the Kun Commune of 1919 (#29), the site of the illegal communist party printing office in the 1920s (#36), and the fact that a certain "martyr"[13] in 1956 worked in a particular factory (#27). Most of the plaques are marble or stone with a simple collection of words and refer to people even most Hungarians today could not readily identify.

---

11 Items in the Statue Park collection are identified by numbers in parentheses in the text. These correspond to the numbers on the plinths and the numbers in the catalogue (see Statue Park n.d., a).

12 In 2006, this was the equivalent of $2.84 each.

13 Those who fought and died on the side of the government during the 1956 revolution were referred to as "*mártirok*," the martyrs.

**Figure 7.5 "The Endless Parade of Personalities of the Workers' Movement,"
The Statue Park**

This is the case for several of the figures. In spite of the fact that Ede Chlepkó (#26) was important enough to not only be sculpted in a bust but to have a square in the XIX district named after him, when asked, the thirty-something-year-old guide on the Hammer and Sickle tour admitted that he had "not heard of him really ... they had to use unknown people." While the tour guide added that Chlepkó was, he thought, a district leader, *Utcák, Terek, Emberek* (FSZEK 1973), a toponymic guide to the streets and squares of Budapest, indicates that he was active in the Hungarian communist party during the period of the First World War and the Kun Commune, he emigrated to the Soviet Union in the 1920s and was imprisoned under false charges in 1938, dying in captivity. The printed catalogue for the park gives a similar story, though without the accusation of false charges and places his arrival in the Soviet Union and his death a year later (Statue Park n.d., a). Although he was closely aligned with the 1919 Commune he has nevertheless been relegated to a small paragraph in both documents and is poorly recognized by Hungarians today, unlike his more famous counterpart Béla Kun.

As the leader of the first communist government in Hungary during the aftermath of the First World War, Kun was a natural subject, however controversial, for communist hagiography. He is present in three separate monuments including one of the largest pieces in the park, which originally had been erected in 1986 in the *Vérmezö* [Blood Meadow], a long, narrow park which sits at the foot of Buda Castle (#24) (see Figure 7.6). It was here, the story of Deszö Kosztolányi alleges, that Kun was seen fleeing Budapest in an airplane when the communist uprising was crushed in 1919. He flew low over *Krisztina Város* as this area is called, and those on the ground could see that he was absconding with *Zserbó* pastries in his pocket and treasures from his aristocratic lovers, one of which, an *aranylánc*, a gold chain, fell as the plane ascended higher and was found later by a *régi krisztinai polgár*, a longtime citizen of the area (Kosztolányi 1963). A more prosaic explanation for the placement of the statue is that Kun was said to have addressed the workers' regiments in 1919 in the *Vérmező* as they were leaving for battle (Boros 2002).

Figure 7.6 "The Béla Kun Monument" by Imre Varga, The Statue Park

The monument, made by Imre Varga, consists of a series of flat figures in bronze and chrome steel, proceeding en mass in the direction of the extended arm of Béla Kun who stands above them on a small dais that intentionally evokes a gallows. Next to Kun, sitting slightly askew is a lamppost that shone day and night when the statue was in the *Vérmező*. Kun died in the Soviet Union, a victim to false accusations and the party purging of the 1930s. The monument was, according to the sculptor who offered his interpretation in the 1990s, well after the change of regime, a critique, a problematization of Kun's place in history and hence of the communists in general; "a grimace which ought to belong as our national self-portrait"(Varga in Prohászka 2004, 210). Thus the lamppost, which in Hungarian literature has come to symbolize hanging was, according to the artist, a symbol of where Kun should have ended up (Varga in Prohászka 2004, 210), and rather than the leader directing fighters, Kun should be seen as a man condemned to face the ghosts of those for whose deaths he was responsible (Varga in Boros 2002, 35). The complexity of the sculpture's message is lost to those visitors relying only upon the official guide book. It narrates the passage from Kosztolányi's story of Kun's flight, the official favor the artist was kept in, and the fact that the piece, before it was removed from the *Vérmező*, was "wrapped" and painted by the political opposition (Statue Park n.d., a). In spite of

Kun's role in Hungarian communist history, many of the foreign visitors to the park will be unaware of his importance.

Even international figures such as Dimitrov (#14, #15), the head of the Bulgarian communist party, are little appreciated by visitors who have in mind the big names like Lenin and Stalin.[14] Among the collection are a number of pieces which represent general themes imposed by Soviet ideological aims from the time of the Red Army's victory all the way down through the 1970s. These included the celebration of Soviet-Hungarian friendship, gratitude for Soviet liberation at the end of the war, the honoring of Soviet war heroes, and, of course, the glorification of the worker. Many of these are legible on the surface, but, as with any work of art, require a careful reading of the iconography and sometimes inscriptions to get a fuller picture.

The Hungarian-Soviet Friendship Memorial by Zsigmond Kisfaludi Strobl (#4) shows a man in a button down shirt and pants, shaking hands with a soldier in a heavy overcoat and fur cap. In order to read this image one must be able to recognize the ubiquitous "uniform" of both the proletariat and the Red Army soldier. Furthermore, one must notice that the worker grasps the soldier's hands with both of his, while the soldier retains his left hand in a stoic fist. The significance of the statue, depicting the utter gratitude of the worker and the unequal relationship which obtained between Hungary and the Soviet Union was not lost on those who toppled the statue during the 1956 Revolution nor upon the communist leadership which quickly had it replaced (Boros 2002; Prohászka 2004). The question remains whether it is grasped by those visiting the park today.

The removal of these statues from their original locations has implications for understanding the significance that they had in the everyday urban fabric of Budapest (James 1999). Unlike the House of Terror which has the task of making manifest what was known but unmarked, the Statue Park has the task of resignifying statues, plaques, and monuments that not only lost meaning with the change of regime, but which also lost their original intent when removed from the places designated by those commissioning the works and those who undertook the design of these pieces. The most ridiculous examples are plaques which are now installed on brick facades, plaques which say *e házban* [in this house], *e gyárban*, [in this factory].

Yet, as those who envisioned this park understood, the removal of these pieces from the public spaces of the city was the best way to ensure their continued existence as didactic tools (Boros 2002). The graffiti on the Liberation Monument by István Kiss (#5) bears witness to the threat posed by these ideologically charged pieces remaining in situ. In some cases, monuments could simply be rehabilitated by the removal of offending portions, the Liberation Monument on *Gellért* Hill offering the perfect example. Without the problematic Red Army soldier (which found its way to the Statue Park (#3)) and communist star, and with a simple reworking of inscription, it has metamorphosized into a well loved symbol of the city, which can now be found on t-shirts sold to tourists.

---

14 There are in fact three pieces which depict Lenin; the statue which is in the entrance façade (#1), a Plaque (#13) which served as a toponymic device for a stretch of the Grand Boulevard, and a full size statue (#17) which stood outside the Csepel Iron & Metal Works until it was privatized. There are no images of Stalin.

In many ways, the collection at the Statue Park seems to offer itself up for understanding even without a nuanced reading of iconography or communist history, which is why it makes such an effective appeal to foreign visitors. On the surface the meaning of the park is easily accessible; "here are images, they are overdone, this was communism." There is a sense of irony, or even humor that anyone would believe that these symbols could have any rhetorical power over a population, that they could be effective tools of repression. This is especially the case with some of the more unfortunate pieces in the park, including: the Soviet Heroic Memorial by Barna Megyeri (#12) which has developed an awkwardly placed rust stain across the heroic soldier's crotch; the grotesquely stitched together colossal hands holding aloft a marble sphere that is a Monument of the Workers' Movement (#31); and the overwrought Martyrs Monument (#38) depicting the last moments of a dying victim, which many visitors are inspired to reenact for their colleagues' cameras.

The reality is that the presence of these monuments within the landscape served to remind residents of the control exercised by the Communists and this domination of space was only the most visually obvious. How to remind people of this long after the offending statues had been removed was a question that concerned Eleőd, the architect who designed the layout of the Statue Park. "Not to scorn-to remind" he states (Boros 2002, back cover).

> I tried to consider every aspect of the task with the seriousness it deserved. Of course, I cannot claim to know what the ultimate Truth is. There is meditation. There is time. I had to realize that if I were to make this park with more direct, more drastic, more up-to-date, methods (as others thought fitting), if, that is, I were to turn these propaganda statues into a theme park of counter-propaganda, then I would have done nothing but follow the prescriptions and accept the mentality bequeathed to us by the dictatorship. [Boros 2002, 6]

What Eleőd has created is a subtle critique of communism that depends heavily upon a precise understanding of the way in which the communist regime dominated space and the way in which these tropes have been carefully referenced by the architect. Even for Hungarians versed in the duplicitous language of communism, the symbolism can be opaque. An understanding of the meaning behind Eleőd's organization of space within the park depends heavily on the textual gloss provided by the guidebook and without it the pieces, stripped from their original locals, can tends towards kitsch or, even worse, miscellaneous irrelevance.

It is the manipulation of space in Eleőd's plan that gives coherence and meaning to the pieces in the Park. This plan draws together the façade entrance and the end wall linked through the "one true path." On either side of this axis are three figure-eight paths, each echoing a trope within the socialist movement's repertoire of propaganda and which help to organize the disparate elements of the collection—"The Endless Parade of Liberation Monuments," "The Endless Parade of Personalities of the Workers' Movement," and "The Unending Promenade of Workers' Movement Concepts." In the midst of the "one true path" lies the flower bed laid out to display the Five Pointed Red Star of the Communist Party.[15] This planter recalls a similar

---

15 The display of this symbol is actually illegal in Hungary except for artistic purposes.

prominent Red Star flowerbed which sat between Castle Hill and the Chain Bridge on the Buda side of the city.

The overgrown bedraggled nature of the flowerbed, an unintended consequence of the lack of funding for the Statue Park, lends an air of decrepitude to these monuments, as does the unpaved, unlined parking lot which stands in front of the main entrance. Many of the plinths have suffered cracks in the cement and some have lost bricks which have not been replaced. The chain link fence which demarcates the park from its environs also lends an aura of temporality. It is only the main façade wall and the end wall which close the park in with bricks.

Both of these walls are pregnant with symbolic meaning about the nature of communist rule. The main entrance was built as an intentional façade, a "Potemkin wall" in the grand style of Socialist-Realist architecture (Boros 2002, 11). The main gate with its 200 lines from Gyula Illyés' *One Sentence on Tyranny* is forever shut and it is only through the side entrance, a reference by the architect to the back door dealings which were required to get things accomplished under communism, that access can be gained. The emptiness of the promise elicited by the façade is made apparent; there is nothing behind but a little ramshackle building housing the ticket office.

The terminal wall offers an implicit critique of communism as an ideology. This wall, a line of red brick, is simply there, embedded into the ground at an angle and a place which ignores the actual topography, the reality on the ground. The guide book explains the symbolism:

> ... the path which has guided the visitor so far, goes past the two statues of the negotiators, which have become a symbol of "farewell", and then a few meters further, travels smack into the end wall. You can progress no further. You have to turn back. It is a dead end. [Statue Park n.d., a]

Without consulting the text, the end is apparent, though the awkward alignment of the wall seems a case of sloppy building. The wall, as well as the rest of the space within the park, is an empty signifier without the text. Eleőd's plan is complex and opaque without explanation.

The poem which appears on the front gate gives direct insight into the way in which the designers of the park thought about communism and although it is translated in the guide book, it is questionable whether visitors read it. Illyés' poem, which was published in the midst of the 1956 Revolution, asks where to seek out tyranny and answers that in fact it is everywhere, in a wife's kiss, in the mother's smile, in the plate and glass, in the snow, in the priest's sermon, in monuments and the painter's brush; everyone and everything is implicated (Illyés 1989). It offers a more nuanced and more complicated rendering of the dictatorship of communism than that told by the House of Terror which is able to lay the blame upon foreign armies and advisors and a select group of complicit individuals. For those involved with the Statue Park, communism was something that happened to everyone and thus the best way to move forward into democracy is to talk, to show, to educate, to remind. As Eleőd commented, "On the gates of Statue Park, the poem by Gyula Illyés ... was conceived with the same intention–among the statues, the same silence rumbles as in the poem. Agony, mourning, impotence, shame, consternation, rage and spite ..." (Statue Park n.d., a). Because of this humility in not claiming to know the truth of the period, the Statue Park's story may be lost in the openness of the design.

Conclusion

The end of communism in Hungary brought about sweeping changes in all aspects of life, including a major reshaping of the urban landscape within Budapest. The signs and symbols of the former regime were swept away; names were changed and offensive statues and monuments were removed. At the same time, the space opened up by democracy allowed for new signs and symbols related to communism to emerge, though these, rather than celebrating the heroes of the party and the workers' movement, brought attention to the crimes and victims of the regime. In some cases, the celebratory was simply replaced by the accusatory as in the case of the raising of a crucifix and sign recalling the 1951 destruction of the *Regnum Marianum* church in the *Városliget* after the giant *Fegyverbe* [to arms] statue, which had itself replaced the church, was removed to the Statue Park.

The debate over how best to deal with the period of real existing socialism engaged all areas of society from art to economics, from health care to politics. As Michalski (1998) notes, the removal of offending statues in the aftermath of 1989 lacked the pathos and intensity of the toppling of the Stalin statue at the height of the 1956 Revolution. Instead, most statues, plaques, and monuments were simply taken away by order of the local governments. Granted, some monuments were marred with graffiti, but for the most part, the complex tasks of dealing with a new economy and a new political system overrode the need for massive demonstrations of iconoclasm. In some cases, statues were removed in spite of their being fondly received by the general population, as in the case with Captains Steinmetz (#40) and Ostapenko (#41).

The Statue Park and the House of Terror present two differing versions of the way in which communism should be remembered. They are two contenders in what K. Horváth (2005) terms a "battle over socialism in the symbolic field." While both offer critiques of the former regime, the extent to which blame can be neatly attributed, and the ease by which the narrative can be written are at the heart of these two different approaches. The House of Terror exists in situ, the site of actual torture and terror, while the Statue Park is exiled to the outskirts of the city. Yet both sites have had to find ways to re-inscribe meaning upon place and upon objects in place. The House of Terror uses all instruments at its disposal, from "representative" tables which are like tables that some dissident might have sat at, to music and lighting reminiscent of a Hollywood movie. The visitor is led through the doleful tale, by the careful manipulation of the spaces of the former villa, as well as the narrative devices arranged in a tableaux of blame. Reconstruction always treads a thin line between truth and fantasy (Eco 1986) and the House of Terror does not shy away from embellishing the truth when the physical reality is lacking, as in the reconstructed prison cells. While little is left for the imagination, the visitor leaves with a sense of having seen first hand the facts of real existing socialism. For the most part, Hungarians who visit the House of Terror leave satisfied since what they have seen fits with their versions of the events, and for those who have a different set of memories, they simply do not go.

The Statue Park offers yet another means for narrating the past, one based on the stripped down presentation of material cultural artifacts. The details of the story must be filled in by the visitor. The catalogue helps in this task, but for those eschewing guidance, the narrative logic which underlies the juxtaposition of the bust of Endre Ságvári and the statue of Árpád Szakasits remains obscure. The task which the designers of the Statue Park set forth was a difficult one, to remember an ideologically overdetermined time in an ideologically free manner. The repository of items in the Statue Park are genuine artifacts of the regime but stripped from their places and the raw power which supported the ideology that birthed them, they are dependent upon a reinscribing of meaning by drawing a line around them which says "here are relics of the past which we have saved from the oblivion of time; pay attention as these tell a story." The line, though clear, is itself endowed with meaning by the architect, a meaning which helps to shape this disparate collection, but a meaning which depends heavily on textual gloss. And so the question remains as to how effective a mnemonic device the Statue Park will be since, as it stands, without further funding it is relegated to the periphery spatially and socially, while the House of Terror sits in the full view of the denizens and visitors to Budapest making manifest what was once forgotten in the space of the city and offering itself up as an easily digested chunk of historical narrative. In the end however, what may be the most lasting memory of communism will not be found in either of these two sites, but rather in the still functioning, still flourishing panel flat estates that ring the city and can be seen by any who care to look around.

Chapter 8

The Parallax of Landscape: Situating Celaque National Park, Honduras

Benjamin F. Timms

Introduction

Landscape has a long and distinguished history within the discipline of geography, but by its own admission is but one mode of ordering the facts of the geographical world. The origins of landscape within geography are traditionally traced to the nineteenth century German ideal of *Landschaft*, with a dual meaning of (1) an area of land and (2) the viewpoint of this area of land tied to the perspective of an individual from a particular location (Olwig 1996). A dualism exists in the definition in that the landscape is both the material reality of a physical area of land *and* the ideal representation perceived from a particular viewpoint. However, it is acknowledged that landscapes can be viewed from multiple perspectives, each giving a differing interpretation (Meinig 1979).

Parallax, the apparent change in appearance of an object when viewed from different positions, is conventionally associated with the discipline of astronomy and, within geography, in the practice of photogrammetry. However, the definition fits nicely with the tradition of landscape geography where perspective is of the utmost importance. Hence, is it possible to create a complementary parallax of perspectives to address the human–nature dichotomy imbedded in the creation and global diffusion of that staple of tourism, the national park? Here this question is addressed by situating Celaque National Park, Honduras, within differing perspectives of the landscape tradition. It is argued that the ideological separation of humans and nature that occurs within the American national park tradition must be addressed if the negative cultural and natural repercussions of this exclusionary model of national parks are to be resolved.

National Parks as Landscape

As a national symbolic place, Yellowstone was perceived as a representation of national ideals and national self-image: a grand landscape painting that stood as a metaphor for America itself. [Germic 2001, 9]

A national park is formally defined as a bounded physical area of land protected from human habitation and exploitation for the conservation goals of biodiversity protection, continuation of ecological services, and other more spiritual, intrinsic, and aesthetic values (McNeely 1990; Furze, De Lacy, and Birckhead 1996; Mitchell 2003). National parks are also significant destinations for tourists. However, a brief overview of the current literature regarding national parks from the landscape tradition challenges the official view that national parks are purely natural areas without human influence, and hence in need of protection from humans. In contrast, both directly and indirectly, they are cultural landscapes (Germic 2001).

The origins of the formal national park, with notable antecedents in forest and game reserves (Harrison 1992; Wright and Mattson 1996), arose in the United States out of the nineteenth century assertion of environmental consciousness by prominent writers such as Henry David Thoreau, George Perkins Marsh, and John Muir. Guided by these influences, and in opposition to the English cultural landscape of well-manicured parks, America placed cultural value in the scenic landscape beauty of untamed wilderness, "… it was obvious that what America lacked in cultural treasures it more than made up for in natural wonders. The American landscape became an effective substitute for a missing national tradition and a repository of national pride" (Patin 1999, 41).

In addition to the naturalist writers, who often wrote the description of nature through cultural imagery such as "Devil's Tower" and "Montezuma's Castle" (Patin 1999, 42), the American school of landscape painting contributed to the creation of national identity out of the valuation of nature. It was in this tradition that the majestic natural landscape was depicted through prose and paintings that created a symbolic landscape through which cultural meanings and values were encoded (Patin 1999). The existence of a general consensus among Americans of a certain vision of natural landscapes is in part due to indoctrination through painting, poetry, literature, and landscape design (Nuemann 1998). Frederick Law Olmstead, the father of American park landscape design, goes so far as to offer natural landscape appreciation as a measure of an individual's civility (Grusin 1998).

Toward the end of the nineteenth century the creation of the first national parks in the majestic wild lands of America began in earnest with the establishment of Yosemite and Yellowstone National Parks. However, these parks were both areas of "land protected from human exploitation and occupation" (Brechin et al. 1991, 7), as idealized by the National Park Service, and substantially influenced by the American urban park and landscape architecture tradition of Olmstead (Carr 1998). Thus, the Yosemite landscape attempts to both preserve and protect nature, and provide for the recreation and recuperation of those who labor (Grusin 1998). Through careful deployment of the principles of landscape architecture, Yosemite was made accessible to tourists, while maintaining the feeling of wildness. The end result is a reproduction of nature in Yosemite for human consumption through the practice of tourism.

National parks as realized in the United States are analogous to art galleries and museums (Patin 1999). In a technique similar to that applied in landscape painting, nature becomes an object to be displayed through the design of scenic roadways, exhibits, and lookouts which all regulate visitor movements to maximize the visual presentation of natural landscapes. These design features reflect the classical landscape perspective through the removal of evidence of human labor and the separation of the observer from the land. As such, national park landscapes relate to the German idea of *Landschaft* with a dual meaning of (1) an area of land, and (2) viewpoint of this area of land, tied to the perspective of an individual from a particular place (Olwig 1996).

As can be seen, the general assumption that national parks are for the conservation of nature becomes more complicated when viewed through the lens of landscape geography. For example, the creation of national parks arise out of a geography of exclusion (Mitchell 2001); "On the one hand, the reservation of 'nature' required and facilitated reservations for various Native nations" (Germic 2001, 9) and as a result, the history of the conquest of Native American societies was transformed into the conquest of nature. Both Yosemite and Yellowstone first had to be emptied of their native inhabitants before they could become "natural."

Today the geography of exclusion inherent in protected natural areas has diffused across the globe. In 1962, the first list of protected areas contained a mere 1,000 sites. By 2003 it had grown to over 100,000 protected areas covering approximately 11.5 per cent of the earth's land surface, roughly the size of the South American continent (Chape et al. 2003). However, the dichotomy present in exclusionary principles based on the American experience of national parks has also spread and, as a result, recent estimates place dislocated populations in the tens of millions globally (Brockington, Igoe, and Schmidt-Soltau 2006). As is clearly evident, the repercussions of the human–nature dichotomy are not inconsequential as this new class of environmental refugees would attest (Geisler and de Sousa 2001).

Protected areas such as national parks ignore the historical, and often contemporary, reality of these "natural" landscapes. For example, creating islands of biodiversity in parks where attempts to preserve a static nature from change is paradoxical in that nature is forever changing as a dynamic system (Pletsch 1993). The static exclusionary environmental strategy which believes nature can be "located, fixed, and preserved outside of culture … read[s] generations of social actors out of the 'nature' they preserve, denying any social history of landscape" (Katz 1998, 55). Few, if any, natural landscapes are without a social history (Stevens 1997; Redford, Robinson, and Adams 2006). All natural areas protected as parks have been altered by humans to some degree and denial of this "illustrate[s] man's short memory where his own effects on landscape are concerned" (Nelson and Byrne 1966, 226).

It is argued here that the negative repercussions of national parks, from social impoverishment to unintended environmental degradation, are a direct result of the denial of the human-nature relationship. As this brief overview of landscape perspectives on national parks reveals, they are indeed cultural landscapes. But to get at the heart of the matter we must delve deeper into the topic, and Celaque National Park in the western highlands of Honduras will be used as a case in point to do so. By situating this national park within several traditions from landscape geography, insights gained can help provide a new parallax vision of parks that attempts to address the human-nature dichotomy.

The Morphology of Celaque National Park

> The cultural landscape is fashioned from a natural landscape by a culture group. Culture is the agent, the natural area is the medium, the cultural landscape the result. [Sauer 1925, 343]

Carl Sauer's (1925) landmark work *The Morphology of Landscape* brought the German *Landschaft* ideal to American geography and serves as a basic template for modern traditions within landscape geography. Sauer's morphological method first describes the physical properties of an area and then how these are modified by the human agent of culture into a cultural landscape. The analysis of Celaque National Park that follows similarly begins with a brief description of the physical properties of the area of study.

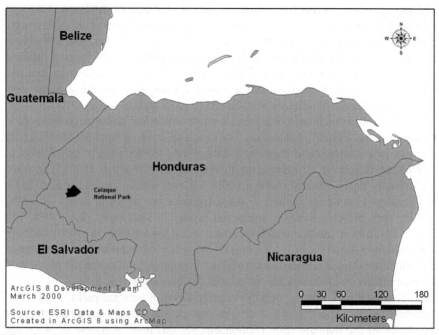

Map 8.1 Celaque National Park, Honduras

The Cultural Landscape of the Celaque Mountains

The Celaque Mountains, located in the western highlands of Honduras, have their geologic origins in an ancient volcanic plateau of the Central American volcanic axis (see Map 8.1). Due to their volcanic genesis, the soils are highly fertile. This contrasts with much of Honduras, where the infertile Old Antillia geologic base is dominant (Portillo 1997). The rugged topography of the Celaque Mountains includes the highest point in Honduras at *Cerro de las Minas*, lower-elevation pine/oak forests, and the largest cloud forest in Honduras at higher elevations. High levels of precipitation source dozens of rivers which serves as the origin of the *Lencan* name for the mountains, Celaque, meaning "box of water" (AFE-COHDEFOR, GTZ and GFA TERRA 2002).

Due to the fertile soils and abundant precipitation, the highland cloud forest has a unique assemblage of flora including broadleaf trees reaching 25–30 m in height, epiphytic mosses, orchids, and bromeliads (Meija 1993). Further, high profile fauna species such as quetzals, toucans, spider monkeys, pumas, and at least 45 endemic species of reptiles and amphibians reside within the park (Wilson and McCranie 2004). Cloud forests are considered to be among the most threatened ecological systems in the world, making the importance of the Celaque Mountains a high priority for the conservation of abundant and unique biological diversity (WCMC 1998; GEF 1999; GTZ 2002).

As in Sauer's definition, the cultural landscape of the Celaque Mountains is inherently related to the physical landscape. While the vast majority of Honduras has been geologically endowed with poor soils and low levels of pre-Colombian indigenous habitation, the relatively fertile soils of the western highlands has facilitated higher densities of the indigenous *Lenca* people. Even though the *Lenca* are the largest indigenous group in Honduras in terms of numbers, they are one of the least understood peoples in Central America. The loss of the *Lenca* language, which had completely disappeared except for a few words by 1900, precipitated a decline in traditional *Lenca* culture. Today only a relic of the pre-Hispanic *Lenca* culture survives in place names and syncretic mixes of Catholic and *Lenca* belief systems (AFE-COHDEFOR, GTZ, and GTA TERRA 2002).

Figure 8.1 Traditional Lenca Waddle-daub Dwelling with Fired Wood in the Highlands of Celaque National Park

The *Lenca* of the Celaque Mountains have traditionally resided in dispersed homesteads loosely organized into settlements, or *aldeas*. The dispersed homesteads consist of one-room homes made of wattle-daub walls and pitched roofs of thatch in lower elevations and fired wood shingles in higher elevations where large trees are available in the oak/pine forests and at the highest elevations in the cloud forest (West 1958). The homesteads are surrounded by agricultural fields, which in turn separate homesteads from each other (see Figure 8.1). The resulting *aldeas* are not centrally organized settlements but a dispersed association of homesteads whose patterns follow the contours of mountain valleys.

The highland *Lenca* rely primarily on subsistence agriculture for their livelihoods, using a traditional swidden fallow system (Jansen 1998; Brady 2003). They produce a particular assortment of crops and handicrafts in accordance with the physical environment. The traditional *milpa*, a Central American agricultural system of intercropping maize and beans, is practiced throughout the Celaque Mountains. However, in the higher elevations unique climatic and soil conditions allow for the cultivation of additional crops such as cabbage, potatoes, and carrots as well as the production of ceramic pottery for sale at market. The pattern of agricultural and handicraft production reflects the altitudinal life zones present in the high relief of Central America.

The assemblage of crop varieties and handicrafts are exchanged at traditional *Lenca* market centers optimally located at an intermediary location between the high and low elevation *Lenca aldeas*. Brady (2003) reports the preference of these markets over larger nearby *ladino* market centers, which are largely avoided by the *Lenca*. In the Celaque Mountains, this can be observed in the *Lenca* Sunday market in the municipality of *Belen Gualcho* on the western side of Celaque National Park, where residents of *aldeas* in the Celaque Mountains exchange produce and crafts with those from lower elevations and purchase basic household goods.

The natural landscape and *Lenca* cultural group have combined to form a unique cultural landscape that has managed to exist for several hundred years, with inevitable adjustments over time as no cultural landscape is immune to change. For example, encroachment by *ladino* farmers and export oriented coffee estates have created a geo-cultural pattern of verticality with the traditional subsistence *Lenca* cultural landscape in the upper reaches of the Celaque Mountains and *ladino* farms and coffee estates in the lower elevations and surrounding areas. Yet the basic symbiotic cultural landscape has been maintained in the face of ever-encroaching outside pressures for change which has spanned the historical spectrum from the Spanish conquest to the repressive Honduran state. However, this cultural landscape has recently been threatened by the contemporary ascendance of the national park landscape over the older symbiotic cultural landscape of the Celaque Mountains.

Celaque National Park and the Disembodying of the Cultural Landscape

In 1987, the Celaque Mountains became another type of cultural landscape, a national park. The park was originally created to protect sources of water through the conservation of highland forests (Herlihy 1997; Portillo 1997). But in contrast to the American national park model, resident populations were allowed to remain within the park's boundaries so long as no further forest clearance occurred (AFE-COHDEFOR, GTZ, and GFA TERRA 2002).

During the 1990s, Honduras was forced to adopt sweeping neo-liberal adjustment programs imposed by international financial organizations (Jansen 2000). One aspect of these programs was the ceding of protected area funding and management to non-governmental (NGO) interests (Beltrán and Esser 1999). As a result, control of Celaque National Park was transferred in 1997 to the NGO *Proyecto Celaque*, funded and created by the German development agency GTZ (GTZ 2002). *Proyecto Celaque*, in line with the American national park landscape perspective, created a park management plan calling for the relocation of *Lenca* peasants residing within the park, a reversal of the original Honduran policy.

In October of 1998, just as *Proyecto Celaque* took over management of Celaque National Park, Hurricane Mitch struck Honduras, leaving widespread devastation throughout the country. In the wake of destruction a *Proyecto Celaque* report described the situation as an "*oportunidad*" to implement the newly construed policy of relocation of the resident *Lenca* (Oviedo 1999). In response, and in keeping with the new policies, requests for assistance by the affected *Lenca* peasants residing within the park were denied unless they relocated to settlements outside the park's boundaries. Approximately half of the *Lenca* residents of Celaque National Park "voluntarily" relocated in order to receive assistance, albeit under extreme duress (Oviedo 1999). While the proximate cause of relocation was a natural disaster, the ultimate cause was the implementation of exclusionary Celaque National Park policies predicated on an American ideal of a natural landscape devoid of human habitation.

The end result of this brief morphological study is that the landscape of Celaque National Park contains a varied cultural history. It is not a purely natural landscape, but contains several cultural imprints that have modified the landscape in differing ways. Hence, the application of cultural landscape traditions to the study of the park is more appropriate than treating it as a solely natural landscape, as the national park perspective attempts to do. But to ask why this has not been acknowledged requires the application of additional landscape traditions to understand how the ideological representation of the landscape can be responsible for the practical treatment of it.

The Insider versus Outsider Landscape Perspectives

In the early 1970s, a reactionary movement to the quantification revolution arose within geography, termed "Humanistic Geography." The movement had a basis in the philosopher Søren Kierkegaard's subjective humanism where the only truth is in the individual. David Lowenthal (1972) and Yi-Fu Tuan (1974a&b) were at the heart of this tradition, and both discuss how our representation of a landscape is more about what is in our own heads than what is in the empirical landscape.

All humans share the physiological senses and a world view centered upon them. From that point on, however, experiences deviate to create a unique world view for the individual. Culture plays an important part as we are taught, or imprinted, by our cultural group to place importance on certain aspects and even which senses we rely upon. Further down the hierarchy, kinship and family diverges our perspectives on reality in a similar manner. Finally, one ends with the individual, who has a unique view of the world based on personal experience, biological faculties, and accumulated values. As a consequence, each individual will have a matchless perspective and interpretation of landscape.

Tuan (1974a) discusses the creation of an insider versus outsider view of landscape. The insider is a person who is actually part of the landscape. Their world view, which influences their perspective on the landscape, is shaped through the intertwined relationship of livelihood. On the other hand is the perspective of the outsider, a person who is not from, or part of, the landscape. This worldview is based on cultural and personal experiences created elsewhere, and hence the ability to intimately read or know a landscape in other settings is encumbered. The outsider perspective can only be ephemerally based on the superficial. As an example, take the landscape of a national park. The outsider may view it as a natural landscape without human influence. However, the insider may be quite surprised to find out he or she resides in, and depends on, a landscape that is defined as having no human influence.

For Celaque National Park, the outsider perspective is embedded within a "western" cultural view of natural landscapes. Western conceptions of nature are centered on the separation of humans and nature (Stilgoe 1982), which predicates the perspective of national parks as (or as they should be) devoid of human influence. However, the insider perspective of Celaque National Park belongs to the residents within the "natural" landscape who are reliant on it for their very livelihood and location of their homes. The inherent duality, then, is that Celaque National Park is a natural landscape devoid of humans as an ideal *and* a cultural landscape where humans and nature are inherently intertwined as a reality.

"I love the mountains, but I hate the park."
– *Lenca* peasant relocated from Celaque National Park

The Landscape of Social Formation

Cosgrove (1984; 1985) focuses not just on how we perceive landscapes, but how the purposes of social formation brought on by capitalism are reflected in landscape representation. He traces the idea of landscape back to the Italian Renaissance painters of the sixteenth and seventeenth centuries who took Euclid's geometry and applied a linear single-point perspective to represent a three-dimensional landscape in two dimensional paintings with mathematical certainty. The painter, or observer, would represent the landscape according to the way they viewed the world, in effect appropriating and controlling the representation of it. Commonly this representation of the landscape would remove all evidence of the peasantry with a brushstroke and replace it with an idyllic scene for the appreciation of the elite.

More importantly, though, is the new perspective represented, both through landscape painting and the related surveyor's "malicious craft" (Cosgrove 1985, 55), the legitimization of the existing social order by representing space as something that could be controlled and dominated as an object transformed into the private property of an individual or state (Cosgrove 1984; Barnes and Gregory 1997). The political economy of social relations appears in how landscape is represented by a new way of viewing coincident with the infant stages of an emerging capitalism. During the Italian Renaissance, rich merchants from the mercantile era were able to appropriate country lands for leisure, and in later eras, profit.

Cosgrove (1985) addresses Tuan's (1974a) idea of the difference in perspective between insider and outsider. According to Cosgrove, the traditional insider is not separated from the landscape, but connected to it through the relationship of a natural economy. However, outsiders are separated from the landscape, and in the era of an emerging capitalism, such as occurred in Renaissance Italy or during the Enclosure Movement in England, an alienated relationship arises between owner and commodity. Such an idea can be traced to Marxist theory which claims that primitive accumulation violently separated humans and nature through the enclosure of lands and common property, turning them into private property, in effect making land a commodity (Sessions 1991).

Cosgrove's ideas are evident in the concept of national parks, defined as a natural landscape devoid of human habitation and exploitation (Davey 1998). Just as an artist would clear away any existence of human toil and labor in creating an idyllic scene, the human history of national parks must also be erased. The outsider's perspective stemming from the Western capitalist culture of Europe and the United States dominates as the landscape becomes not just an ideal representation of pure nature, but a real physical appropriation of the landscape from the insider through the formal creation of the park, marginalizing the insider perspective in the process.

Examples abound of national parks dispossessing the insider, from the violent history of park creation in the United States where Native Americans were forcefully removed (Germic 2001) to similar cases throughout Africa (Grove 1990; Neumann 1998; Cock and Fig 2000). The conservation of these areas has resulted in the transfer of ownership of resource riches but marginally located landscapes from local insiders to external outsiders (Smethurst 2000). Further, in relation to the perceived need to remove human labor and presence from natural landscapes, specifically in the form of relocation, national parks can be seen as creating aesthetic landscapes for philanthropic causes and tourist consumption (McNeely 1990; Shafer 1990; Castree and Braun 1998; Escobar 1998; Katz 1998; Zimmerer 2000). The rural poor are seen as a liability, and hence are removed from the scene as the landscape becomes appropriated by relatively powerful outsiders.

For Celaque National Park, the designation of the land as a national park under the control of the state between 1987 and 1998 appropriated the landscape from the *Lenca* residents, albeit tempered with the allowance of regulated continued settlement within the park boundaries. But the bequeathing of park control to international interests in 1998 resulted in the relocation of residents from their highland communal *aldeas* to the private land tenure settlements outside of the park, which is reminiscent of the capitalist transformation of natural economies tied to the representation of landscapes. Further, one of the major arguments for creation of the park resides in the tourist revenues it is expected to create, supporting the view that economic concerns play an important role in social representations of landscape.

Organic versus Universal Landscape

Olwig (1996; 2002) acknowledges Cosgrove's interpretation of the role of the Italian Renaissance painters' contributions to the formation of landscape, but delves deeper into history by examining the concept of landscape in the German speaking lands of northern Europe. Basing landscape on the German concept of *Landschaft*, Olwig discusses how the term originally did not just mean a bounded area of land and the view of it from a single point observer. *Landschaft*, based on other derivates of the term in related languages such as Dutch, included customary organic laws that bound the community together through social norms and also bound them to the landscape in a nexus of localized land and culture. Such customary organic laws differed from location to location based on the local environment and local culture. However, forces of unification within larger regions encompassing multiple *landschaften* had to destroy localized customary organic laws binding communal communities to the territory in order to impose a more universal legal law that bound all people to the state.

Figure 8.2 Relocated Lencan Child Walking Along the Outside Boundary of Celaque National Park

Customary organic laws defining the relationship between the community and the landscape have to be severed in order for universal laws to apply in creating a National Park landscape. Through official state designation of the Celaque Mountains as Celaque National Park, and the subsequent ceding of park management to outsiders, universal laws and regulations relating to the western concept of national parks override local customary organic laws of *Lenca* communities within the park. When the original allowance of residence within the park was replaced by the policy of resident relocation, the local *Lenca* communities were physically severed from the land upon which their culture depended (see Figure 8.2).

Landscape as Social Compromise

Mitchell (1994) moves beyond the landscape tradition of Cosgrove and Olwig by proposing that landscape does not just *reflect* cultural landscapes, but also has an element of *intent*. Drawing on Marxist dialectics, he claims the creation of one landscape reflexively creates alternate landscapes. For example, the creation of Celaque National Park and the implantation of its exclusionary policies created an ideal "natural" landscape while necessitating new settlements outside of the park for the recently dispossessed. The separate landscapes differ, yet revolve around the same issue of appropriation of landscape. In the process the landscape is contested, and so landscape becomes a social compromise between opposing social groups.

Duncan (1993) discusses the need to address "the other," those left out, in representing or creating landscapes as well as the importance of treating landscape at broader spatial scales. The majority of landscape geography is tied to specific localities, relatively small in area, which can be attributed to the single observer perspective of *Landschaft*. But in today's globalizing world there are forces at grander spatial scales that influence the local.

Mitchell (2001) addresses such forces in relation to the contestation over the control of landscapes wherein insiders fight to wrest control from globalizing forces which, in turn, seek to homogenize the world through the exertion of power and influence at all scales from global to local. There is intent involved, and contestation, all of which take place within and about landscape. Mitchell and Duncan, by expanding the influences on local landscapes to global processes, adopt elements of political ecology where multiple scale analyses are needed to uncover all forces involved in the creation, alteration, and contestation of landscapes.

For Celaque National Park the questions of contestation are readily apparent, and are best exhibited by the resident *Lenca* populations originally being allowed to remain in existing settlements within the park, albeit with regulations. The later relocation of residents also exhibits social compromise in the creation of two settlements located in differing geographical locations. The modified park regulations and alternate relocation strategies represent contestation over landscape, intentionally creating alternate landscapes of social compromise.

One settlement, *Otolaca*, was created by NGOs, state agencies, and the Catholic Church on land far removed from their original highland *aldeas*, in effect severing the *Lencan's* connections to their cultural landscape. However, in the other settlement of *Los Horcones* the relocated *Lenca* were able to negotiate the process whereby in exchange for promising not to return to reside within the park they were allowed to determine their own resettlement site. As a result, they chose a location in geographical proximity to their previous *aldeas*, but outside the park's boundaries. This amount of self-determination allowed them to exploit a loophole in national forest polices and as a result any land continuously kept cleared of trees remains in their ownership (Jones 1990). Since the *Lenca* were geographically situated closer to their original landholdings within the park, they were able to work both the land outside the park and within it (see Figure 8.3). In effect, they resided outside of the park, keeping their side of the bargain, yet were able to expand their land holdings and exploit different ecological niches present at different altitudes.

Figure 8.3 Lencan Settlement Within Celaque National Park

Conclusion

Parallax, the noticeable change in appearance when an object is viewed from different positions, is clearly discernible in the various yet complementary landscape perspectives surrounding national parks. This chapter began with the formal view of national parks as "natural" landscapes devoid of human influence. But this idealized single-point perspective masks the multiple realities that exist within the landscape, which were only uncovered when applied through the multiple lenses of traditions within landscape geography applied here.

For Celaque National Park, the application of these landscape concepts shows how singularly defined natural landscapes can in fact be cultural landscapes created out of differences in insider/outsider views, alterations to livelihoods and land relations, and social compromise arising from contestation. The landscape is in fact not one landscape, but a combination of several landscapes depending on perspective. The difficulties facing national parks, whether in the United States or other localities such as Honduras, revolve around this disconnect.

Whether addressing national parks or urban-based landscapes, there are an infinite number of valid landscape perspectives. While the idea of an infinite amount of perspectives conjures up fears of a downward spiral into post-modern chaos, there are avenues to build outward. The concept of parallax can be one such solution. When applied in stereoscopic photogrammetry, the simultaneous viewing of photographs taken from complementary positions creates three-dimensional perspectives with the ability to perceive depth. Can such an opportunity exist for landscape? If so, landscape must move beyond the geometry of a linear single-point perspective that transforms the three-dimensional reality into a two-dimensional representation of it. Perhaps a complementary parallax of landscape perspectives can form a basis for just such a movement.

The American national park perspective of natural landscapes devoid of human impact or occupation is premised upon the separation of humans and nature, ideologically and practically. However, most natural areas protected as parks have been changed appreciably by humans. The very idea of protecting nature from human impact through the creation of preserved landscapes is inherently paradoxical and reflects the philosophical dilemma created by the perception of humans as separate from nature. This is especially the case when the preserved landscapes are created at least partly for touring, as in the American national park tradition. The path to resolve this dilemma is through a re-conceptualization of the human/nature dichotomy.

The inherent unity of humans and nature must be accepted if there are to be steps taken toward the conservation of sustainable human-nature systems. In the Celaque Mountains this relationship was not acknowledged and the presence of *Lenca* peasants within park boundaries was assumed to cause environmental degradation. However, this assumption is challenged by research showing that, in fact, forests within Celaque National Park actually increased between the year of the park's creation, 1987, and the year of relocation, 1998 (Aguilar 2003). During the same time period additional research has found evidence that the expansion of coffee production is encroaching into the park itself, forming a significant threat to the ecological goals of the park (Southworth, Nagendra, and Tucker 2002; Southworth, Tucker and Munroe 2002; Bass 2006). To compound the issue, all of the relocated *Lencan* households have been relegated to working as increasingly cheap laborers on the same coffee estates in order to make ends meet, in effect facilitating the expansion of coffee cultivation into the park. Thus, the implementation of an exclusionary landscape based on the ideological perspective of nature as separate from humans has created pressure on the landscape the park was designed to alleviate, while impoverishing the local population at the same time.

In light of these revelations, can it be possible that the cultural landscape of the *Lenca* could serve as a buffer against the expansion of capitalist export agricultural production, the true threat to the conservation goals of the park? Viewing the landscape from the perspective of a complementary parallax of culture and nature addresses the human/nature dichotomy, a dilemma which has plagued national parks.

> For those of us who are convinced that landscapes mirror and landscapes matter, that they tell us much about the values we hold and at the same time affect the quality of the lives we lead, there is ever the need for wiser conversations about ideas and impressions and concerns relating to the landscapes we share. [Meinig 1979, 47]

Chapter 9

Insiders and Outsiders in Thy

Daniel C. Knudsen[1]

Introduction

The purpose of this chapter is to examine tourism in Thy, a region in the northwest of Denmark. This examination is grounded in the literature on landscape studies with particular reference to the meaning of the landscape of Thy to "insiders" and "outsiders" (Lowenthal 1972; Tuan 1972, 1974a&b). The term landscape is taken here to mean the outcome of successive years of human–environment interaction that marks a place as someplace. For the purposes of this chapter, one can think of "insiders" as those who live in Thy on a permanent basis and "outsiders" as those who visit the region as tourists.

By pursuing this approach, two things are accomplished. First, by grounding tourism studies in the social theoretic approaches to landscape, this chapter explores both the ways in which tourist sites are constructed and the ways that tourist landscapes are filled with intended and unintended meaning for the tourist, thereby improving the current theorization of tourism (Ooi 2003; Hollinshead 1994; Urry 1990; 1992a&b; Rojek and Urry 1997). Second, within the tourism landscape, this chapter explores the insider and outsider viewpoints. These two stereotypical viewpoints allow a clearer understanding of the landscape upon which tourism takes place.

The theoretical points of this chapter will be grounded with reference to Thy, an area of approximately 1,000 square kilometers in northwest Jutland (see Map 9.1). It is bordered on the west and north by the North Sea, on the south by the *Limfjord* and on the east by a former fjord that has since silted-in to form Europe's largest bird sanctuary. Known for its long summer days punctuated by both sun and driving rain, its steep bluffs facing the North Sea, and rocky beaches, Thy is the heart of the "Danish Riviera." Natural beauty and historical significance combine in Thy to provide a landscape of quite different meaning to insiders and outsiders. For insiders, Thy is filled with reminders of cultural and historical significance, while for outsiders understanding the historical and cultural significance of Thy is difficult (see Chapter 1).

1   I would like to acknowledge funding from the West European National Resource Center and Office of International Programs at Indiana University. This research could not have been possible without the translation assistance of Ms. Pia Tripsen.

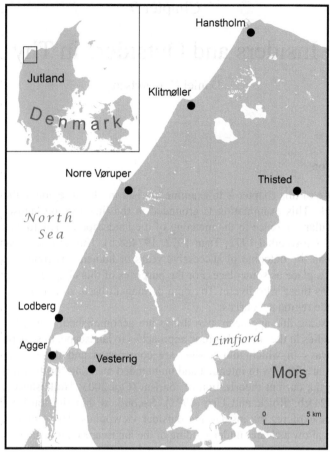

Map 9.1 Thy, Denmark (by Shanon Donnelly)

In what follows, the literature on landscape is examined. The chapter then focuses explicitly on the literature of "insideness" and "outsideness," a literature that draws heavily from semiotics; it then analyzes the tourist landscape of Thy, contrasting outsider and insider views of the landscape with the help of excerpts from interviews conducted in the area in 2005.[2] The chapter closes with a discussion of the theoretical and practical consequences of this analysis.

Insider versus Outsider

Like Lowenthal and Tuan, this chapter uses the terms "insider" and "outsider" to contrast those who can read the landscape's meaning in a certain way and those who

---

2   In what follows, interviewees are denoted by number only to preserve anonymity. For example, source #1 is denoted as (S1).

cannot. Thus the point is to emphasize that, in any view of the landscape, there is a multiplicity of insider and outsider meanings. In using this approach, the chapter is situated clearly within the various social theory approaches to tourism. Davis (2001, 126) notes the importance of social theory for providing a "deeper understanding of landscape processes and their significance" as well as its importance for explaining how tourists behave and why.

The distinction between insider and outsider has been considered largely within the framework of the way in which the two archetypical groups relate to the landscape (Green 2005). Aasbø (1999, 149) notes that:

> The insiders tended to talk about the view as seen *from* their private places, while outsiders viewed the landscape as from a point outside it ... The difference is here one of respectively seeing out of and into the landscape, of being a participator in it and being a spectator to it.

The distinction is key because to an insider, a landscape symbolizes a daily pattern of "work, social contact, daily experience and other pragmatic criteria" (Aasbø 1999, 149). On the other hand, outsiders employ "visual and physical descriptions" having their basis in "ready-made academic systems of categorization" (Aasbø 1999, 149). The comparison of insider and outsider viewpoints is thus always one of the "social and the personal versus the physical and the visual" (Aasbø 1999, 149).

At the heart of the difference in insider and outsider views of the landscape lie differences in lived experience with the particular landscape in question. To understand the origins of this difference, Peircian semiotics is useful. Unlike Saussure who viewed the interpretive process as diatic—the mediation of signifier (word or thought) and signified (object) through social convention—Peirce viewed the interpretive process as triadic (Eco 1995). In Peirce's case the mediation of signifier and signified is affected by what Peirce refers to as "collateral information"— lived experience brought to bear on the interpretive moment (Uslucan 2004). The importance of this can be seen immediately from this simple example. For Saussure, all persons subjected to the same social conventions, (for example, all Danes), will interpret a given thing the same way, while to Peirce that is clearly not the case since all persons (Danes) do not share the same lived experience.

The importance of this concept for this discussion is that, on the one hand, insiders can each clearly have differing interpretations of the landscape "as seen from their private places," but more importantly, outsiders also see the landscape from their own "private places"—places that are far removed from those of the insiders and that are, generally more derivative of "ready-made academic categorizations." This occurs because in the absence of knowledge about a specific landscape, one seeks to interpret landscape in reference to what he or she has experienced (Heckenberger et al. 2003; MacCannell 1986).

Typically, the outsider is seen as "writing the insider into nature" (Pipkin 2003, 10). Thus, it is typical to configure the outsider/insider relationship as a power relation, wherein the outsider wields power over the insider. Such arguments are, for example, at the root of Urry's (1990) conceptualizations of the tourist gaze and are central to post-colonial studies. In connection with the latter case, colonialists are in a position to willfully use the power and resources at their command to rearrange the physical landscape to suit their purposes, thus altering irrevocably the cultural landscape.

For example, Slater (1993) examines the manner in which the English landlord class of Ireland willfully created the picturesque in the countryside of County Wicklow's landscape through both addition of plant materials, many of them exotics, and fake ruins, and the removal of items that might spoil the vistas whether they be single cottages or whole villages. These additions and subtractions to the cultural landscape are important precisely because, by controlling the landscape, the landlord class was able to (re)create history and thereby potentially alter the memory of the place as held by insiders (Crump 1999).

Thy

In contrasting the views of "outsiders" with those of "insiders" when it comes to Thy, it is useful to consider Thy in transect from south to north, the way it is most likely to be experienced by an outsider. Viewed in this fashion, the experience of Thy begins at the *Limfjord*, continues northward across the farm area to the limestone bluffs, across the bluffs and downward through first the forest and then the heath to the fishing towns and the beaches of the North Sea coast. Each of these locales offers a contrast between outsider and insider views.

If Thy is a journey from *Limfjord* to the North Sea coast for outsiders, for insiders, Thy is a complex and utterly nostalgic landscape that contains all of Danish history within its symbols. For the insiders quoted below, Thy is a place with a unique character:

> For me it's special. [S1] It is not found anywhere else in Denmark. [S2] It's the most varied nature in the same spot [in Denmark]. This coastline as we have here, you won't find it elsewhere. It is a rough nature. Thy has a special nature ... There is no other place where you have these dunes and no houses for a long time. [S3]

The Limfjord

For outsiders, the *Limfjord* offers a low-cost alternative to summer on the North Sea. Populated by holiday camps, it offers limited sailing and wind surfing opportunities. According to the author's own (outsider) children, the small towns of the *Limfjord* are an intensely boring place for children on vacation, although they may be the best spot to actually interact with Danes, who themselves are not on holiday.

The insider knows the *Limfjord* is the central east-west road of Danish history and commerce. With medieval Thisted on the western end, prehistoric Viborg perched at the center and medieval Aalborg on the eastern end, the *Limfjord* has been the center of North Sea and Baltic trade since Neolithic times. It is from bases along the *Limfjord* that Viking longboats sailed to the coasts of England and France. It is from the *Limfjord* that Knud the Great invaded England, establishing the *Danelaw*. As the Danish crown's power diminished, trade along the *Limfjord* and with Norway continued to be important and the homes of Danish captains who plied the *Limfjord* are collected on the various *Skippergade* [captains' streets] that nestle in the towns that dot the edge of the *Limfjord*. Today, the cities of the *Limfjord* serve as global centers of agriculture, engineering and electronics. Thisted retains its role as a center for global agricultural exports, while Viborg maintains its previous political clout as the principle administrative center of north Jutland, and Aalborg remains the pre-eminent engineering and electronics capital of north Jutland (Sørensen 1997). However, the *Limfjord* is more than history. It is a set of relationships with neighbors, of stories told at family get-togethers and it is the feeling and smell of living adjacent to an arm of the sea.

The Land Between

The land between the *Limfjord* and the limestone bluffs that form the backbone of Thy offers little to outsiders in the way of tourist attraction, outside of the occasional Neolithic passage grave in the middle of a farmer's field. Filled with farms, it is an area for outsiders, not unlike the Great Plains of the United States, an area simply to be traversed in movements between the major commercial and transportation centers of Thy which sit on the *Limfjord* and the North Sea resorts to the west and north (see Figure 9.1).

Figure 9.1 A Passage Grave in a Farm Field

For insiders, the area between the *Limfjord* and limestone bluffs is quintessentially Denmark—"the nation of farmers." Populated chiefly by hog farms, the landscape contains reminders of the ancient Danish estates (both relic and newly-constructed). Freedom from manoralism was granted only in 1788, and landed versus peasant status remains a (largely unspoken) part of Danish identity to this day. At the same time the modern hog farms are reminders of the post-manoralist growth of the highly successful Danish cooperative movement that has its origin in Thisted and which spread from there first throughout Thy, and then to the rest of Denmark.

Beyond the freedom from serfdom and the silent testimony to Danish cooperation lies a history of struggle to make the land productive. This struggle pitted the inhabitants of Thy against the sea and drifting sands to the west and north on the one hand, and the national government over the reclamation and subsequent preservation of land on the other. To residents of this area farming has been a proud struggle against both nature and the central government. Evidence of the struggle with nature is embedded in their everyday lives:

> [You should see the Tvorup] church ruins. I think they moved them there in 1785. It was the second time they had to move the church inland to save it from the sand flows. They also had to move the Vang church. Before 1900, the sand was always a problem. In the 1800s, they blamed the sand on the digging of the English Channel. As the sand got worse, the big farms moved inland to get away from the sand. There used to be sand storms. Sand covered everything. There was a bad storm in 1938. It destroyed fields. This led to the new farm insurance law for crops in 1938. After the storm the dikes were several meters wider. Every year at Easter, we would get a wind for three days. These don't happen any more. In the 1930s, the Heath Society started the shelter belts around each house. Bigger farms were not good about donating land to help plant trees. [S4]

Once the problems with drifting sand were solved, turning the heath into productive agricultural land was a further challenge:

> It is hard to change heath to farmland. Big ditches have to be dug and you have to turn the soil upside down so that you put the sand under the soil. Then we planted mainly oats and barley. That was 50 years ago now. Today, we no longer bother. [S4]

Struggle with the central government has two aspects. The first centers on a strong sense of the peripheral nature of the place with respect to the government in Copenhagen, mixed with the underlying belief that Teutons (as natives of Thy refer to themselves) are far more intelligent than widely believed:

> But mainly the rest of the country doesn't care about Thy. We are kind of a periphery with those kind of problems. They know our industries and our mink farms, but maybe that is all. But here in Thy, we think more than we talk. [S4]

The second has much to do with preservation policies in the areas north and west of the farmland, many of which have been incorporated into National Park Thy.

> Now there are more deer. They come out of the forest ... They eat our maize, so we have stopped growing it. There are also many swans. They are a real problem for us. Last year they destroyed a whole field of rapeseed. They are very clever and they live a long time, so we may have trouble for many years. There is no crop insurance for animals, only hail

insurance. In 1992 we got a little money from the government because it was such a dry year and we only had 20 per cent of our normal yield. [S5]

And, as in any farm community there is the memory of the past configuration of agricultural practice. Says one farming elder of the changes seen in his lifetime:

> Every house was a farm when I was young. Today there are only two to three farms left [here]. There has been lots of farm consolidation using money from the bank. Those that borrowed money and bought more land, they survived. [S5]

> Years ago, after the big manors, there were cottagers, but now again there are big fields. The little farms don't pay enough and it is more mechanized. We started with horses, but it became more mechanized little by little. I have seen a lot of changes in 80 years. [S4]

The Bluffs

The 100 m elevation limestone bluffs tower above the rest of Thy. Largely barren except for windmills, churches and wind-bent trees, this is the first sublime landscape of Thy, surpassed only by the beach itself and the cliffs at Bjilburg where limestone bluff and sea meet, marking the east end of Thy.

Figure 9.2 Vestervig Church, the Oldest Church in Thy

To insiders, like outsiders, the bluffs are sublime, but in a different way. For insiders days with wind and driving rain are "good Danish weather." And there is no better place than on the bluffs of Thy to experience it. The majority of the churches in Thy also are situated in this quintessentially sublime landscape. Indeed, the churches that populate the limestone bluffs are among the oldest in all of Denmark, some dating from the original conversion in the eleventh century. The churches, because of their age, have seen the entire religious history of Denmark from the original conversion, to the coming of Protestantism, to the nineteenth century revivals of Grundtvig and the Home Mission, to modern, largely secular society (see Figure 9.2).

The churches and their surrounding cemeteries are not the only references to the sublime. The seaward foot of the bluffs contains the simple stone markers denoting the last resting places of the Viking forbearers of the modern Danes. Juxtaposed against this ancient tableau are the fields of modern Danish windmills that also populate the limestone bluffs—the pride of Danish engineering that account for a little over 11 per cent of the country's energy needs.

The Dunes and Coastal Towns

The area that lies between the limestone bluffs and the coast is comprised of vast stretches of reclaimed coastal dune. Populated by forests, lakes, peat bogs, expanses of heather and, as one nears the sea, the coastal towns, it is the most anonymous for the outsider and the most significant for the insider.

For the outsider, this area is a place for recreation—camping in the forest, golf in the clearing, trekking in the forest and the heath. The coastal towns are the focus of activities in the evening and on the frequent rainy days. Strolling the breakwaters and harbor roads, purchasing souvenirs and curios in the shops, visiting the museums at Agger, Lodbjerg, Norre Voruper, or Hantsholm, and dining out are principal pastimes of outsiders.

For the insider, the area has a rich history. In the Neolithic period, both the dead and living sacrifices were placed into the bogs. Subsequent excavation has established the pre-history of what today is Denmark. The forests in Thy are large plantations that date from the middle of the nineteenth century when efforts to reclaim land lost to dune encroachment finally took hold. The heath that occurs coastward from the forests is similarly important for the role it plays in stabilizing the area between the forests and the carefully planted marram grass of the primary dunes in this created landscape. As with the farmland lying between the *Limfjord* and the bluffs, these seemingly natural areas are fraught with struggle against nature and central authority, and are sources of local pride for having triumphed over nature. The forests and heath have a complex history:

> They planted marram grass first on the dunes. They started the forests with little mountain pines and then afterward planted different trees. The Vang plantation burned in 1938. Forest workers replanted it. They were local forest workers. Now they start with scrub oak. [S4] They have introduced new kinds of trees now. They plant more broadleaf trees. The foresters tell me the broadleaf trees don't work here, but they are under pressure from Copenhagen, so they do it even though they don't grow well here. They have been testing

broadleaf trees in the forest here [Tvorup]. They are planting oak. [Recently, the heath] has increased on its own because of new regulations. [S5] [Before then], the heath had not increased much in the last 50-100 years. [S4]

The coastal towns are reminders of the close ties Danes have with the sea. From these coastal towns the Danish fishing fleet once sailed the North Sea, which has now become so depleted and polluted that fishing has seriously declined and only a few smaller vessels leave the beaches each day to venture out to sea. Still, closeness to the sea is essential to Danish identity—an important point of national pride is that no place is more than 30 minutes' drive from the sea (see Figure 9.3).

Figure 9.3 Klitmøller, a Typical Coastal Town in Thy

In the towns, fishing is still embedded in the memory of those living there:

Earlier here in Thy, it was very poor. It was just fish and bread. [Today] there are more tourists, before there was a lot of fishing here. When we were kids, they were still pulling the 14 *tonne* boats up here. The landing place was the life-blood of the community. There were basins for the lobsters that were caught and you would go down and buy them. Klitmøller exported fish. The holiday houses were just for people who came here from Thisted. [S1 and S3]

After the war there were 25 boats here then and three or four men worked each boat. They fished with mussels as bait. The wives made the lines ready for fishing. They use a long line, 300-500 meters in length, with hooks every two meters. They caught cod, catfish and *langefish*. They also caught plaice in the autumn. Lobsters were trapped in summer. Each boat had 300-400 lobster boxes. [S3]

People lived by fishing in Hantsholm and by working in the fishing factories. The last few years there has been a depletion of fish in the North Sea, and so there are fewer jobs for these people. Fishing is not good. You get money from the government when you take a boat out of the fishing fleet. Except for the area near Norway, the North Sea is shared by the EU member states and the Dutch use heavy nets that drag the sea bottom, even in the rocky areas, and destroy the sea bottom so nothing can live. We Danes complain, but then they try to restrict our use of young herrings for industrial use. So we can't do anything. [S3]

The Beach

The beach is, of course, the central focus for outsiders and it is the reason that they come to Thy. Whether strolling along the beach examining shells or hunting for amber in the driving rain or sunning on the warm summer days, it is the sublime feeling of exposure to nature that is the principal draw of "the Danish Riviera." For those outsiders who have returned more than once, there is a familiarity with the restaurants and shops, with certain cottages and perhaps the coincidental friendship. There are the known places of fierce undertow. There are the usual trails across and along the primary dune, the haggling for fish and mussels when the boats return from the sea at noon. And most of all, there is the leisurely lying about.

While for the insider, there are these experiences, but also a few more. First, there is the subtle enmity between insiders and outsiders that plays out spatially—on the beach, among the houses, in the restaurants, in the bars, and in the shops. "I don't like all these tourists. The German tourists don't follow the laws," intones one local (S1). There is the pastiche that comes with tourism that is somewhat regrettable from the insider point of view—the Bornholm fish smokery, for example, plopped down in northern Jutland as if it had been there all along.

And there are the hulking reminders of history—the relic Nazi bunkers used as restrooms and changing houses when they have not slid all the way onto the beach. While these are arguably reminders of World War II for outsiders and insiders alike, few features of the landscape of Thy better represent the complexity that geographical scale introduces to meaning in the landscape. Reactions range from curiosity to paradox to grieving and remembrance:

Fishing was limited during the war. Three boats were lost to mines during the war. Many children were fatherless. [S3]

Ultimately, it is precisely at the beach that life in Thy is measured by the insider. "The sea has taken from the beach every year," says one elder insider remembering a childhood when the beach was 2 m wider with a knowing nod, as if his own life is measured against the endless hunger of the sea for the land (S3). In other parts of Thy, insiders claim it has swallowed up whole villages in the last 200–300 years (S5).

Conclusion

This chapter has used notions of "inside" and "outside" perspectives (Lowenthal 1972; Tuan 1972, 1974a&b) to illustrate two versions of a tourist landscape. The author is, of course, an outsider, but, through extensive interviews, has gained some insight into insider perception. By unwrapping the meaning of landscape, intellectuals are better able to understand what tourists see and therefore better understand tourism itself. Previous theoretical frameworks, which have leaned very heavily on Foucault's notion of *le regard* largely fail to acknowledge the role of meaning particular to the tourism experience. As a result, such explanations fail to account for much of tourism—that portion that is "unguided."

In Thy, the contrasting meanings of places to insiders and outsiders is palpable. Insiders see the *Limfjord* as the east-west road of Danish history, the land that borders the *Limfjord* as inscribed with the history of Danish agriculture, the bluffs as the locus where God and life connect, and the heath and sea as enigmatic of the long struggle with nature for survival. To them Thy is filled with a deep symbolism. The outsider struggles to decipher Thy's symbols, if one bothers to try at all. What is to be made of streets named "*Skippergade*," farm fields with burial mounds, churches on bluffs constructed from granite stones, forests where the trees are planted in rows amongst sandy hillocks, and relic bunkers used as changing rooms?

In the absence of deeper understanding, tourists, as outsiders, fall back on the sublime and romantic aesthetic of landscape. A detailed knowledge of history and culture is required to interpret landscape in any full or rich way. One must realize this in explanations of tourism and this in turn means that cultural/heritage tourism has limits—limits that are bounded more by acculturation than by tourism policy or infrastructural expenditure.

Meaning is far from given in tourism. As is the case with all objects and places, it is contested through discourse and carefully negotiated. Geographical scale is critical to the construction of meaning in that meaning is transitory across scale. Objects and place have both "insider" and "outsider" meanings and each of these two forms of meaning do not preclude the other. This is so because in the absence of "insider" meaning, "outsider" meanings are constructed for what one gazes upon by making reference to available concepts. For Westerners, these are the aesthetic and romantic devices of the late eighteenth century (Zaring 1977). It is through this process, that places and objects, however alien and unfamiliar they may be, are rendered interpretable.

Chapter 10

Tourism as a Reconnection to the Neolithic Past: The Tamgaly Rock Paintings of Kazakhstan

Altynai Yespembetova, Jillian M. Rickly, and Lisa C. Braverman

Introduction

This chapter will contrast the insider and outsider gazes within current tourism theory by focusing on the area of Tamgaly in Kazakhstan, a tourism site known for its Neolithic petroglyphs. Urry (2002, 12) argues that a distinctive tourist gaze is based on "certain aspects of the place to be visited which distinguish it from what is conventionally encountered in everyday life," suggesting several ways in which this distinction "between the ordinary and the extraordinary is established and sustained." The first category involves "absolutely distinct objects to be gazed upon which everyone knows about" (Urry 2002, 12). In reference to Tamgaly, this category of the gaze is much more established for the insider, the native Kazakh, as a sacred heritage site. By analyzing the insider meaning of Tamgaly, one can understand the historical value of the connections between Kazakhs today and their ancestors in this sacred landscape. This category of the tourist gaze commonly "entail[s] a kind of pilgrimage to a sacred centre" (Urry 2002, 12). Urry's second category, gazing upon a particular sign or type of landscape, is much more characteristic of outsider perspectives of Tamgaly in which the site represents a type of landscape in which the petroglyphs, the rock paintings themselves, are the objects of the gaze.

These categories are further supported by Soper, Knudsen, and Yespembetova (2003), who argue that the interpretability of tourist sites is generally based on the formation of a dominant discourse concerning the contested meaning of those objects and places. Once established and made hegemonic, however, meaning is geographically contextual. Geographical context enters into the consideration of theory of tourism in two ways. "First, meanings of objects and places can be geographically pandemic or endemic. Second, and in a related way, meaning shifts across geographical scale. For example, that object deeply understood endemically may have only a vague meaning or even be meaningless pandemically" (Soper, Knudsen and Yespembetova 2003, 3). Zaring (1977) reminds one that, in the absence of insider meaning, we fall back onto the more general meaning of objects and places in our experience. Individuals have their own understanding of their surroundings, which forms different levels of perception (see also Davis 2005).

The analysis provided by this chapter on differences in insider and outsider gazes is based on Meinig's (1979) classic paper "The Beholding Eye: Ten Versions of the Same Scene." By examining the gazes of insiders and outsiders in this fashion, the major dimensions of the difference in gazes can be emphasized.

Tamgaly

The landscape of Tamglay, with its Neolithic petroglyphs, is a common site of heritage tourism for the country of Kazakhstan. Within Kazakhstan there are 17 petroglyph sites, situated mostly in the southern, eastern, and central regions. Tamgaly, however, covers a larger area and contains a denser concentration of different types of petroglyphs than the other sites.

Besides the petroglyphs, the landscape of Tamgaly as a whole is symbolic for Kazakh culture. Set within a canyon, Tamgaly contains high rock faces, springs, and grassy valleys, as well as other sacred landscape features, including burial mounds and religious altars. For the nomadic ancestral Kazakhs, this was a prized location that connected the important aspects of life from subsistence practices to spirituality.

However, to most outsiders, Tamgaly represents the long ancestry of art-making in human existence, of the life of earlier sojourners in the land, and of the land itself. To the outsider, the site has meaning only through the sublime connection between the present and the Neolithic (see Zaring 1977). In what follows, the Tamgaly landscape will be examined from multiple perspectives, with a continual emphasis on the landscape as a heritage site and as a sacred landscape.

Four Versions of the Same Scene

An examination of the Tamgaly landscape, with particular attention to the petroglyphs, can be undertaken using ideas from Meinig's (1979) paper "The Beholding Eye: Ten Versions of the Same Scene." In this paper, Meinig argues that landscapes are comprised both of what we see, as well as our perception of what we see—reality filtered through lenses of lived experience. As a result, when looking at a landscape, we read into it our beliefs, values, hopes, and fears. He continues on by listing ten possible aspects of a scene; landscape as nature, habitat, artifact, system, problem, wealth, ideology, history, place, and aesthetic, and illustrates what each viewer draws upon in interpreting the landscape in such a way.

Of Meinig's ten versions, three are used here: history, place, and ideology. In this analysis, a fourth perception not found in Meinig, national identity, is also used. National identity is a dimension of landscape study, and is especially valuable for insider versus outsider interpretation of the Tamgaly landscape.

History

History is the most important element of the Tamgaly landscape. As a consequence, this section of the chapter will begin by giving a chronological overview of Tamgaly, then discuss the overarching implications of history on the values of place, gazing, and insider and outsider meaning.

The vista that lies before one's eyes, when visiting Tamgaly, depicts a "complex cumulative record of the work of man and nature" (Kadyrbayev 1977; Meinig 1979). It moves minds several millennia back into the history of Kazakhstan. Additionally, Lymer (2004) states that "these sites are significant to Central Asian archaeology as they contain large concentrations of rock art depictions that date as early as the Middle Bronze Age (c.1400 BC[E]) and extend into other later periods up to the present day" (159). As Sauer (1925) argued in his landmark paper "The Morphology of Landscape," the landscape provides evidence of the human imprint created through cultural and historical agency on the physical environment.

Around the second millennium BCE, the climate dried out, lowering the water level in the steppes; one result was that the area became more densely populated (Nikolaevich and Bajpakov 1994). The era that ensued is known as the Bronze Age, during which time many of the rare depictions found at Tamgaly came into being (Nikolaevich and Bajpakov 1994). Though most of the petroglyphs were created in the Bronze Age (around 1400–1000 BCE), some petroglyphs were found to be from the Sako-Scythian period (around 500 BCE), the early Turkic period (around AD 500), and even later (Lymer 2004).

The Saak's nomad era started around the seventh century BCE and formed the "animal" style pictures. The "animal" style art gradually disappears on the edge of the first millennium BCE and the first millennium AD. It refers to numerous scenes of duels between horse warriors and the composition of battles. Petroglyph art degenerates with the penetration of Islam and other religions in Kazakhstan during the medieval period. Petroglyphs dating from the Saak period present the schematic figures of people and skillful pictures of animals and weapons made in the best traditions of nomadic art (Maksimova 1985). Therefore, Tamgaly is an enormously rich store of historical data about the life of Kazakh ancestors and their societies.

Tamgaly as a landscape is a unique place; it represents particular aspects of the historical lifestyle and spirituality of the Kazakh people. Lymer (2001) argues against a common, outsider belief that Tamgaly, along with other petroglyph sites, is merely an "outdoor art gallery" (16). He states, "The key problem with such an approach is that it conceptualizes the petroglyphs as the residual actions of past peoples and this diminishes the role of the images in society to minor afterthoughts" (Lymer 2001, 16). When viewed as an accumulation of past events, peoples, and traditions, the rock paintings themselves offer an artistic link to the past. The "petroglyphs of Tamgaly … were part and parcel of the overall significance of place … actively incorporated into Kazakh engagements with the world around them" (Lymer 2001, 167). However, he notes that "these local perceptions do not simply and mimetically reproduce the given attributes of the place, for present-day practices acknowledge and dynamically interact with the archaeological artifacts of past societies" (Lymer 2004, 167).

When Tamgaly is "gazed upon" by tourists, they most likely glean only an incomplete, outsider version of the site's many meanings. Outsiders will most likely know less about ancient Kazakhs and Kazakh art than current residents of the country; while such a knowledge disparity may not hinder appreciation of Tamgaly, it almost certainly limits outsider identification with the site. Outsiders most likely do not identify immediately with the ancient peoples who created the petroglyphs, but more likely identify the Tamgaly landscape as representative of the numerous petroglyph sites around the world; in situations where this lack of identification is present, an immediate understanding of Tamgaly as a place of heritage is nearly impossible.

Carroll, Zedeño, and Stoffle (2004) note the connection between ritual and place, which is the sacred landscape. They note that, while ethnography, material culture, and geographic indicators are important in the analysis of a ritual landscape, places often become important because rituals occurred there. The notion of connecting a ritualistic past to the petroglyphs of Tamgaly is of considerable value. Tamgaly is considered a possible site of former shamanism (see below); such a past could theoretically give the petroglyphs a ritualistic feel, even at present. Thus, history is inextricably bound to place. While one can separate what precedes the development of a place—what happened during its creation from what that place represents—too little focus on either the historical or site-specific aspects of the place creates an incomplete understanding of the location. As a result any discourse pertaining to Tamgaly's history is tightly intertwined with Tamgaly as a place.

Place

Meinig's (1979) perception of the landscape as a place needs to be transformed into a unique place for the purposes of this chapter. When discussing landscape as place, Meinig states that, "in this view every landscape is a locality, an individual piece in the infinitely varied mosaic of the earth" (Meinig 1979, 45). Thus, in order to study Tamgaly fully, it will be necessary to focus upon the specifics that make Tamgaly distinctive. This work is concerned with the individuality of this landscape. Such a view has to combine geographical and historical approaches and should be directed toward particularity, not generalization. The uniqueness of Tamgaly motivates people to visit it, thus leading to a better understanding of Kazakhstan's heritage and uniqueness.

Unique for its "abundant concentration of images" (Rozwadowski 2004, 55), Tamgaly features a wide variety of human and animal imagery. Particularly, humans and human-like figures are etched near horses and bulls. According to Indo-Iranian folklore, both bulls and horses were used as ancient sacrifices; perhaps this explains why the animals and humans are often depicted close together. Multiple images at Tamgaly feature humans next to, on top of, and under bulls. Additionally, depictions of bulls and horses are often mixed, creating such figures as the "horned horse" and the "bull resembling horse" (Rozwadowski 2004, 55-8). The site is described by UNESCO as having "some 5,000 petroglyphs" (UNESCO 2006). When describing the importance of maintaining Tamgaly as a World Heritage site, UNESCO also states, "Distributed among 48 complexes with associated settlements and burial grounds, they [the petroglyphs] are testimonies to the husbandry, social organization, and rituals of pastoral peoples" (UNESCO 2006).

Much of the human and animal imagery at Tamgaly is noteworthy simply because it is peculiar. When depicting horses, bulls, and humans, the ancient petroglyph artists strayed far from any kind of realistic representation. Humans have circular limbs, tails, and porcupine-like prickly skin/garments, among other attributes. Figures are not just unrealistic, they often undergo marked changes. Tamgaly features a central rock fissure, after which human figures undergo "a fundamental 'metamorphosis' … doubling their size and (featuring) distinctly deeper engraving" (Rozwadowski 2004, 63). Much like the previously described human metamorphosis, select horses at Tamgaly undergo a similar change. Altering in appearance over the course of cracks in the rock, certain horses undergo mane transformations, changes in the way they are viewed (sideways or from behind), and differences in the depth with which they are etched. In the case of one particular horse, as the horse changed, the etchings became shallower (Rozwadowski 2004).

Perhaps more interestingly, Tamgaly's depictions of humans and human graves suggest it may have been used as a shamanistic or religious site (Rozwadowski 2004; Nikolaevich and Bajpakov 1994). The rock paintings themselves feature anthropomorphic figures looking at each other with upraised arms and tails, enlarged heads, and figures with sun-heads (Nikolaevich and Bajpakov 1994; Lymer 2001; Rozwadowski 2004). While the various types of anthropomorphic figures could allude to a variety of shamanistic rituals, Lymer (2001) and Rozwadowski (2004) suggest that the petroglyphs are suggestive of a morphing of solar and human elements. Lymer (2001, 19) suggests that dots which radiate around figures' heads "suggest pulses of light." Similarly, Rozwadowski (2004) explains that the solar cult was a big part of the Indo-Iranian tradition. Therefore, it can be implied that affinities for solar worship were recognized, if not carried out, at Tamgaly. Lymer's (2001, 19) assertion that "the shimmering effects of patinated sandstone rock not only animated the images, but perhaps referred to the power of the place that was amplified with the addition of significant petroglyph images" supports the idea that Tamgaly has held and still holds profound ritualistic meaning as a sacred landscape. Possible implications of this meaning relate, once again, to Kazakh ancestral lineages because Tamgaly is not just a location of early art—it is a burial site as well. More specifically, there are three distinct "burial" sites at Tamgaly. Each includes bones that have been subjected to inhumation, cremation, and cenotaph (Nikolaevich and Bajpakov 1994). Thus, a visit to Tamgaly could logically involve a connection to historical/ancestral memory. For the Kazakh people, such a visit's ancestral implications might make that (ancestral) aspect of a visit more profound or emotional, but does not necessarily rob a large part of the place's meaning for peoples of different heritages.

Ritualistically, Lymer (2004, 161) cites the Sufi tradition of *ziyarat*—"pilgrimage to saints' tombs." As Tamgaly is a site of mass burial and the rock art is so memorable, it is not a far stretch to imagine ancient Kazakhs journeying to the site in respect for the dead; while making absolutely no claim to there being saints buried at Tamgaly, one can still speculate as to the various rituals that took place at the extraordinary site. Certain scholars even believe that Tamgaly was created partly as a "ritual sanctuary" (Lymer 2004, 162). The discourse of the ritualistic immediately gives way to the discourse of sacred landscape. Laden with meaning, the petroglyphs at Tamgaly carve out a distinct ideological space for themselves. Much like Ginsburg's (2005, 281) notion of "embedded aesthetics," it becomes difficult to grasp the full meaning of the petroglyphs if one (theoretically, an outsider) is not familiar with the social intricacies of the surrounding society.

On a deeper level, Tamgaly's uniqueness allows it to stand apart from not only other Kazakh petroglyph sites, but from sites all over the world. Its aesthetic combined with its historical value make Tamgaly into an iconic national image—one that, in fact, possesses its own distinct ideology.

Ideology

To understand landscape as ideology one must see the landscape "elements as clues and the whole scene as a symbol of the values, the governing ideas, the underlying philosophies of a culture" according to Meinig (1979, 42). Meinig argues that landscape as ideology "is a view which clearly insists that if we want to change the landscape in important ways we shall have to change the ideas that have created and sustained what we see" (Meinig 1979, 42). To view the Tamgaly landscape as ideology is especially interesting because of the complex relationship between the Kazakh people and the site; this view incorporates important aspects of Tamgaly as "place" and as "history." It is through this perspective, landscape as ideology, that one can most understand the sacredness of the Tamgaly landscape. However, it is also through this landscape perspective that there is the greatest disparity between the gaze of the insider and the outsider.

Although the petroglyph art of Tamgaly degenerated with the penetration of Islam and other religions into Kazakhstan during the medieval period, the continuation of the dominance of Islam as the major religion of Kazakhstan has not subdued the significance of this place to the people. Sufism, the form of Sunni Islam practiced in Kazakhstan, joins traditional aspects of the religion with an included emphasis on the remembrance and worship of ancestors. Sufism allows for the expression of spirituality outside of the traditional sites of worship associated with the "orthodox" forms of the religion, such as "the practice of domestic rites and pilgrimages to saints' shrines and family cemeteries" (Lymer 2004, 160-161). This aspect of Sufism reaffirms the connection between the present Kazakh nation and Tamgaly's past, therefore allowing this site to function as a place of heritage as well as a sacred landscape (Lymer 2004).

The interaction between present-day Kazakh Muslims and Tamgaly through pilgrimage has been further examined by Lymer, who focuses on the "cultural and socio-political" dynamics of the practice of "tying rags on branches and other amenable surfaces at special places and sacred sites in the landscape" (Lymer 2004, 158-9). This insider practice illustrates the difference between the insider and outsider gaze at Tamgaly. The insider to Tamgaly, the pilgrim, becomes a part of the landscape by tying a rag (*mata*) as a "wish or dedication to the saint of the place" (Lymer 2004, 159). The insider's gaze goes beyond the petroglyphs and towards the spiritual power embodied in the sacred landscape as an ancestral connection. Tamgaly is a multifaceted landscape for the insider, as a gazing point of nomadic livelihood and spirituality. However, for the outsider, Tamgaly functions as a petroglyph site that connects the visitor to a lifestyle unknown and a time past.

Beyond the Islamic connections visible in the Tamgaly landscape, a more nomadic-ancestral, Neolithic connection can also be examined. Shamans, as religious figures of a pre-Islamic, nomadic lifestyle, consistently interacted with nature. They were the religious specialists who communicated with the supernatural for the benefit of an individual or a community. Therefore, Tamgaly symbolizes a "landscape of spirits" where shamans seek lost souls and spirit helpers (Mar´´i`ashev 1994). "[T]he natural rock was not simply a blank surface on which the petroglyphs were engraved; the rock itself was a shiny, active medium and possessed qualities or powers that were deliberately sought out by privileged members of society, particularly shamans" (Lymer 2001, 21). This remains a tradition for modern society, as a physical medium is commonly needed to understand and to mediate thoughts about what is complicated and abstract (Atkins et al. 1998). Lymer (2001) also suggests that the location of the Tamgaly petrogylphs themselves represent the social relations of the community. The Tamgaly figures are "concentrated on a high ledge towards the top of the hill, overlooking the Tamgaly Valley" suggesting possible societal restrictions placed upon site access (Lymer 2001, 20).

Today, as Lymer (2004, 168) notes, "[t]hese holy sites are available to all persons in the community and the specific local character of the sites contributes to communal identity. Therefore, not only does a special place become an important local holy site, but it also becomes a nexus point where the local community, its cultural heritage and their Islamic traditions intertwine, fuse and develop." Tamgaly has special significance as a "clue" to Kazakh culture. The petroglyphs of Tamgaly reflect this landscape and therefore reflect Kazakh culture and identity. Interpreted as metaphor, these images act as symbols for the identity embodied in Tamgaly.

National Identity

Landscape and tourism intersect at national identity, as "both have a profound interest in promoting place uniqueness and differentiation" (Yespembetova 2005, 16). The significance of landscape as national symbol is particularly interesting in the case of Kazakhstan due to its relatively recent independence from the Soviet Union in 1991. As Bunkše (1999, 122) noted in his examination of Latvian postmodern identity, "[f]reedom is particularly troubling for peoples and nations which are emerging from Soviet colonial rule and seeking to define their national identities and finding their place in a world that is undergoing rapid change." Therefore, as Kazakhstan works to develop a unique, "post-colonial identity from its Soviet, Turkic, and Kazakh heritage" (Yespembetova 2005, 16) Tamgaly landscape features will continue to be useful as national symbols. Many of the petroglyphs from Tamgaly represent the emerging national symbols of the Kazakhstan nation-state. For example, the mysterious sun-headed anthropomorphic divinity that possibly relates to the idea of cosmogony, connecting the sun with fertility of the land, is reflected now on the national currency and is the emblem of the international music festival "Voice of Asia." These images as public, national symbols allow the outsider tourist of Tamgaly to connect to the Kazakh insider.

Like landscape as ideology, landscape as national identity also draws on important aspects of landscape such as history and place. As Daniels (1993, 5) argues, "[t]he symbolic activation of time and space, often drawing on religious sentiment, gives shape to the 'imagined community' of the nation" therefore, "any one national identity is always consciously characterized by both a historical and a geographical heritage." For the insider this is one and the same. Kazakhs have a national identity that is geographically contextual to central Asia, and historically and ancestrally a national identity that is tied to the nomadic lifestyle. Their nomadic heritage is one of the very important factors that distinguish them from other central Asian nations, such as the Uzbeks. The symbols of Tamgaly used as national symbols bring both the historical and geographical heritage together by incorporating the images so closely tied to the Kazakh, nomadic ancestry with a distinct location on the landscape. The importance of Tamgaly as a nomadic grazing site, and therefore, the images found there, used as national symbols, helps to reestablish their unique national identity distinct from that under Soviet rule.

The influence of Islam on Kazakh culture did not diminish the importance of the Tamgaly landscape as a spiritually sacred place. Surprisingly, however, the Soviet period also did not diminish Tamgaly's significance as a heritage and spiritual site. Viewed as a "superstitious relic," the Soviet Union enforced anti-religious campaigns against Islam in Kazakhstan (Lymer 2004, 160). Due to the Sufi tradition of religious practice outside of "orthodox" locations, a common practice being pilgrimage (*ziyarat*), "the importance of these shrines was made even greater during the anti-religious policies of Soviet times" (Lymer 2004, 168). The anti-religious campaigns were mainly focused on the "propagation of theology and institutions of urban Islam," although pilgrims to sites, such as Tamgaly, were seen as "subversive" by the Soviets they were not seen as being as great a threat and therefore were less targeted by the campaigns (Lymer 2004, 168). As Lymer (2004, 168) notes, "[d]omestic religious practices went on quietly and became an element of Kazakh culture that survived 'Sovietization,' while their fully nomadic economy and residential patterns of living ... did not." This history further strengthens the symbolic power of the Tamgaly images as images of a surviving Kazakh culture.

However, Kazakhstan today contains a large population of Russians (~30 per cent) from the Soviet Era, which has led the government to declare Kazakhstan a multi-national state and to move the capital from Almaty to Astana. "Most local Russians before independence identified with Kazakhstan primarily territorially, lacking both an ethnic and political content ... Independence has given this identity a political dimension, but the Russians' identification still lacks an ethnic dimension" (Olcott 1995, 289). Therefore, while the national symbols from Tamgaly represent an important cultural link between the Kazakhs and the homeland of their ancestors, they do not represent the multinational cultural complexity of the current population structure, which includes a significant portion of Russian-Kazakhs. However, it is important to note that this chapter does not argue against the significance of Tamgaly's petroglyphs as important symbols of a pre-Soviet Kazakhstan, just that it is equally important to understand the cultural dynamics of contemporary Kazakhstan as a result of Soviet rule. The symbols of Tamgaly do currently represent the national identity of a majority of the population, an identity that was suppressed, and thereby they provide a reconnection to the Neolithic past of Kazakhstan. But what these symbols will mean for future generations of Kazakhs may be very different as this multi-national population evolves.

Conclusion

Tamgaly can be examined on many levels, as a sacred landscape, as a heritage site, and even the petroglyphs alone can be interpreted for their symbolic meaning. This chapter has interpreted Tamgaly's meaning, value, and significance by exploring the common topic of inquiry in tourism studies, the insider versus outsider gaze. Meinig's (1979) study of ten versions of the same scene is especially useful for this analysis, as it focuses on particular aspects that allow this landscape to be seen through the lenses of history, place, and ideology. All three of these versions of the same scene contribute to the fourth way of seeing the landscape, landscape as national identity.

This research discusses the meaning of the Tamgaly landscape both for insiders and outsiders. For insiders, Tamgaly serves as a key to history and culture and emerges as an important site where connections can be made between Kazakhs and their heritage. To the outsider, Tamgaly may appeal as sentimentally sublime in that it provides a connection of modern human beings to those of prehistory. However, the images associated with the Tamgaly landscape connect the insider and outsider because images of petroglyphs from the Tamgaly Valley are used by Kazakhstan to project its national identity and self-image globally. In presenting itself in the way that it wants to be seen by others, Kazakhstan can make a statement to others about the richness of history and culture—that is, "who we are" and "how we want you to see us."

Chapter 11

Landscape, Tourism, and Meaning: A Conclusion

Daniel C. Knudsen, Michelle M. Metro-Roland, and Anne K. Soper

The authors of this volume have offered several different "views" of the intersection between landscape, tourism, and identity, and the way in which these intersect with meaning. Touring occurs in place and the theoretical insights of landscape studies are appropriate for moving tourism theory beyond the notion of "gazing." Specifically, this volume has been concerned with the ways in which landscape, tourism, and meaning intersect. It has involved elaborating on tourism's roles within a broad spectrum of places and events, spanning across time and space, considering different ideologies, varying levels of economic development, and different approaches to tourism.

Summary of the Chapters

The examination of places where landscape, tourism, and meaning intersect has included Strasbourg, France, where the contemporary urban landscape is a tableaux of competing identities. As a territory which has vacillated between French and German control, it is a fitting place for the European Union to set up shop. It is here that a supranational European identity is being given shape in the tangible form of the European Parliament. And it is perhaps also fitting that as national orientation becomes less important in light of the increasing prominence of the EU, Strasbourgians are also claiming a more assertive Alsatian identity. Thus the Strasbourg landscape can truly be said to offer a window into the reification of identity at the local, regional, national, and supranational levels.

We have also seen how landscapes are understood and consequently shaped by different groups operating from different perspectives. In the Copper Canyon region of Mexico, a more rural landscape offers some of the same complexities as seen in the urban context of Europe; change has been affected by all involved in the region from the local inhabitants, indigenous and *Mestizo*, to foreign tourists driving the newly paved road from El Paso. What is apparent is that differing and sometimes contradictory interpretations of landscape have tangible consequences not just for the physical environment but also for the practice of living.

In the Indian Ocean we have seen how the tourist landscape can be used for more than simply generating dollars. In the case of Mauritius, the cultural heritage landscape has become a rich source for creating and fostering a sense of national identity in this multi-ethnic, post-colonial island nation. Tourism is both a tool and an impetus for giving form to the rich historical past that is the shared legacy of the island's diverse inhabitants.

This is also the case, it is argued, in the rebuilding of post-World War II Munich as it turned back to the past and forward to the future in order to disinvest itself of the horrors of the Nazi period. Manipulation of the landscape was a tool to show that in fact Munich was civilized, a worthy member of the "West" and a place that could be enjoyed. The urban landscape became a concrete means to redefine German identity. But, like a palimpsest, history is written into the text which is the landscape, and the Nazi past, in spite of conscious attempts by the state to disavow it, can still be read in the urban cityscape by those who pay attention.

The landscape of the city can be read but so can the landscape of a particular place. In Budapest, Hungary, the experience of "real existing socialism" has been enshrined in two museums which occupy vastly different "spaces" within the urban fabric of the post-socialist city. The communists understood that space could be ideological, and so too do their successors who have used the manipulation of space in these two sites as a potent tool of historical narrative.

Because space is never neutral, it can be contested. We have seen the implications for indigenous peoples in Honduras of importing the American idea of National Park as a pristine, unpopulated, demarcated landscape. This contestation between the indigenous population and foreign managers is the result of an insider view that sees culture, way of life and land as all closely interwoven and an outsider view which perpetuates the myth that landscape is somehow separate and alienable from those who have dwelt in it and have shaped it.

Landscapes, as objects produced by the interaction between nature and culture, between natural forces and history, are open to interpretation, even by outsiders. In the case of Thy, Denmark, the deep and complex symbolism engendered by the *Limfjord* and the heath are fully intelligible only to the autochthonous inhabitants; it is they who see from within. But for the tourists who come to visit, landscape is not so much mute as different since they are not privy to the stories the landscape communicates.

This is also the case of Tamgaly. As a UNESCO World Heritage site, the petroglyphs speak to a common human history. On the one hand their significance transcends the political borders which are drawn around them. As is argued, though, the petroglyphs have a more specific, or endemic, meaning for the members of the Kazakh nation. The paintings are a part of the landscape, a specific landscape in a specific place. Their significance can not be severed from the history of the area, nor from the present day political and cultural milieu in which they are embedded.

Concluding Remarks

When first published in 1990, Urry's *The Tourist Gaze* totally re-theorized the study of touring. The goal of this monograph has been to move tourism theory forward from the conception of *the gaze* by bringing landscape theory into consideration. In the nearly two decades since the publication of Urry's work, tourism as theorized by

Urry has become increasingly problematic to the point that the study of tourism has largely transcended Urry's theorization.

The argument of the authors in this book is that a refocusing of the theory of tourism away from the tourist to the tourism landscape is essential if tourism theory is to move forward because such a refocus allows scholars of tourism to examine the tourism–landscape–identity nexus. Examination of tourism in this way is fruitful because it allows for adequate complexity in studying the relationship between identity and landscape, the relationship between tourism and landscape, and the way in which these intersect with meaning. At the same time, it does not ignore the role of the tourist since landscapes are given meaning both by those who shape them and by those who interpret them. Theorizing tourism in this manner also places the study of tourism squarely within cultural geography and its landscape tradition. Each of these is discussed in more detail below.

The relationship between identity and landscape is complex. First and foremost, this relationship is dynamic, not static, and is thus a historical process (see Chapters 3 and 4). It is also reflexive in that identity is both formed by and forms the landscape (see Chapters 4 and 5). The landscape is similarly an object of intervention by the state in an attempt to shape identity (Chapters 6), or by various actors in order to shape historical narrative (Chapter 7), and it is likewise a forum of contestation (see Chapter 8). Lastly, it is the locus of deep symbolism for inhabitants, a symbolism largely lost on and hidden from outsiders (see Chapters 9 and 10).

Tourism is the act of deciphering the identity of a place from its landscape. As such, it is an extremely complex process since it involves the peeling back of a place's history, layer after layer (see Chapters 3, 4, 9, and 10). It also requires the seeking out of sites willfully hidden (see Chapter 6) or simply neglected in favor of other more palatable (see Chapter 7) or more accessible (see Chapter 5) sites. At other times, it demands that tourists be more aware of the ways in which the Global North impinges upon the priorities of the Global South (see Chapter 8).

Theorizing tourism this way also raises questions about how tourists create meaning from the sights and sites that they visit. Clearly, for the tourist, the outsider, meaning has little or no relationship to identity as it does for the insider, so that it must have other origins. For the tourist, meaning must be gleaned from the materials and actors (guidebooks, guides, short histories, websites) that are part and parcel of the tourism industry as well as the previous experience of the tourists who bring these to bear upon the deciphering of place. Establishing meaning in touring is made difficult by the accretion of layers in place over time (see Chapter 3), by the manipulation of the tourist environment (see Chapters 6 and 7), or by a simple inability to grasp the largely hidden complexity of the landscape (see Chapters 4, 5, 8, 9, and 10). And yet meaning-making does take place.

The Landscape Approach

The landscape literature does not approach interpretation from a single standpoint; it does not offer an omniscient gaze but rather multiple views are afforded by the rich sweep of theoretical approaches found within the history of landscape studies.

Olwig (1996), for example, uses an etymological approach to gauge the actual lived meaning of the term "landscape" to those dwelling in place. Conversely, Cosgrove (1994, 1998) examines the way in which landscape aesthetics, especially those associated with the picturesque, were used to justify the wholesale remaking of the rural landscape in the aftermath of the European enclosure movements (Cosgrove 1984,1998). Lefebvre (1991) demonstrates the ways in which space is not simply a container but rather implicated in the ideological aspects of spatial manipulation. Mitchell (1994; 2003) theorizes landscape as ideologically implicated in power relations. The power of landscape to convey meaning has been well understood by powerful actors who have attempted to shape cities to convey messages of national identity and more recently supranational identity (Atkinson and Cosgrove 1998; Duncan 1990; Olwig 2002).

Furthermore, each landscape is not static but can more precisely be seen as a complex layering of meaning evolving over time. The landscape is not a naïve object with a fixed meaning simply conveyed to those gazing upon it. Rather, tourists and locals both are active participants in meaning making (Tuan 1974a&b), and for each insider and outsider there is a myriad of ways in which a single landscape can be viewed (Meinig 1979). Lastly, by holding together the physical environment with the cultural expression upon the land, the concept of landscape allows that the physical setting of the sites of tourism are as important as the manifestations of culture embedded in these sites.

In searching for a binding framework for the range of societies in this book, it is offered that the application of landscape perspectives can help better understand the social and cultural processes of tourism through which meaning is derived. We recognize that joining landscape theory with tourism theory is just one of many methods that may be used to glean greater understanding, but we have found it offers an analytical framework that has potential in other settings and find that it goes beyond the hegemony thought to be had by tourists and accepts tourists as equal, not privileged, participants in the tourism experience. It also avoids the other tendency in tourism studies of casting aspersion upon the "ugly tourist" who passively and ignorantly consumes sites, because landscapes have meaning, and because they are open to interpretation, even by outsiders.

It has been argued that institutions utilize and structure the tourism landscape in order to convey meaning for tourists to consume, and to promote internal agendas such as remembrance and nation building. Each of these depends upon people coming together in agreement about meaning found in the landscape. However, each individual, whether insider or outsider, tourist or local, will subjectively interpret the landscape, thus bringing a personal, individualized layer of meaning and understanding.

Landscape studies have accepted the role that culture plays in shaping landscape and vice versa. As it is the cultural landscape in its many forms, which is often the object of cultural and heritage tourism, it makes sense to use landscape theory, which takes a comprehensive approach to understanding a site, both from the point of view of the producer, and the consumer. The chapters in this book have exemplified this broad approach to understanding the phenomenon of tourism in place.

Bibliography

A Fővárosi Szabó Ervin Kőnyvtár Budapest Gyűjteményék Munktársai (FSZEK) (1973), *Utcák, Terek, Emberek* (Budapest: Kossuth Kőnyvkiadó).

Aasbø, S. (1999), 'History and Ecology in the Everyday Landscape', *Norsk Geografisk Tidsskrift (Norwegian Journal of Geography)* 53:2, 145-151.

Absolute Walking Tours n.d. *Absolute Walking Tours Brochure* (Budapest: Absolute Walking Tours).

AFE-COHDEFOR, GTZ, and GFA TERRA Systems (2002), *Plan General de Manejo: Parque Nacional Montaña de Celaque* (Santa Rosa de Copan, Honduras: Proyecto Celaque).

Agnew, J., Livingstone, D.N., and Rodgers, A. (1996), *Human Geography: An Essential Anthology* (Oxford, Eng. and Cambridge, MA: Blackwell).

Aguilar, A. (2003), 'Patterns of Forest Regeneration in Celaque National Park, Honduras', UCLA Department of Geography. Unpublished manuscript.

Aitchison, C., Macleod, N.E., and Shaw, S.J. (2001), *Leisure and Tourism Landscapes: Social and Cultural Geographies* (Routledge Publications: London).

Alladin I. (1993), *Economic Miracle in the Indian Ocean* (Stanley, Rose Hill, Mauritius: Editions de l'Ocean Indiene Ltee.).

Anderson, A.E. (1994), 'Ethnic Tourism in the Sierra Tarahumara: A Comparison of Two Raramuri Ejidos', University of Texas at Austin. Dissertation.

Anderson, K.J. (1987), 'The Idea of Chinatown: The Power of Place and Institutional Practice in the Making of a Racial Category', *Annals of the Association of American Geographers* 77:4, 580-98.

Applegate, C. (1999), 'A Europe of Regions: Reflections on the Historiography of Sub-National Places in Modern Times', *American Historical Review* 104:4, 1157-82.

Aschmann, H. (1959), 'The Evolution of a Wild Landscape and its Persistence in Southern California', *Annals of the Association of American Geographers* 49 (Supplement), 34-46.

Atkins, P., Simmons, I., and Roberst, B. (1998), *People, Land and Time: An Historical Introduction to the Relation Between Landscape, Culture and Environment* (London and New York: Arnold).

Atkinson, D. and Cosgrove, D. (1998), 'Urban Rhetoric and Embodied Identities: City, Nation, and Empire at the Vittorio Emanuele II Monument in Rome, 1870-1945', *Annals of the Association of American Geographers* 88:1, 28-49.

Aumeerally, N.L. (2005), '"Tiger in Paradise": Reading Global Mauritius in Shifting Time and Space', *Journal of African Cultural Studies* 17:2, 161-80.

Barnes, T. and Gregory, D. (1997), 'Place and Landscape', in *Reading Human Geography: The Poetics and Politics of Inquiry*, T. Barnes and D. Gregory, (eds). (New York: John Wiley & Sons).

Bass, J.O. (2006), 'Forty Years and More Trees: Land Cover Change and Coffee Production in Honduras', *Southeastern Geographer* 46:1, 51-65.

Beetham, D. and Lord, C. (1998), *Legitimacy and the European Union* (Essex, Eng.: Longman).

Beltrán, J. and Esser, J. (1999), *Protected Area Management: Analysis of the Contribution of Non-Public Sector to In Situ Biodiversity Conservation in Costa Rica, Honduras and Nicaragua, Central America* (Eschborn, Germany: Deutsche Gesellschaft für Technische Zusammenarbeit (GTZ)).

Best, S. and Kellner, D. (1991), *Postmodern Theory: Critical Interrogations* (New York: Guilford Press).

Birks, H.H. et al. (1988), *The Cultural Landscape: Past, Present, and Future* (Cambridge and New York: Cambridge University Press).

Boros, G. (2002), *Statue Park* (Budapest: City Hall).

Boswell, L. (2000), 'From Liberation to Purge Trials in the "Mythic Provinces": Recasting French Identities in Alsace and Lorraine, 1918-1920', *French Historical Studies* 23:1, 129-62.

Brady, S. (2003), 'Guachipilines and Cercos Zanjos: Lenca Land Use in the Guajiquiro Biological Reserve', in *Cultural and Physical Expositions: Geographic Studies in the Southern United States and Latin America*, M.K. Steinberg and P.F. Hudson, (eds.). (Baton Rouge, LA: Geoscience Publications).

Brechin, S.R. et al. (1991), 'Resident Peoples and Protected Areas: A Framework for Inquiry', in *Resident Peoples and National Parks: Social Dilemmas and Strategies in International Conservation*, P.C. West and S.R. Brechin, (eds.). (Tucson, AZ: The University of Arizona Press).

Brockington, D., Igoe, J., and Schmidt-Soltau, K. (2006), 'Conservation, Human Rights, and Poverty Reduction', *Conservation Biology* 20:1, 250-2.

Broek, J. (1932), *The Santa Clara Valley, California: A Study in Landscape Changes* (Utrecht: N.V.A. Oosthoek's Uitgevers).

Bunkše, E.V. (1999), 'Reality or Rural Landscape Symbolism in the Formation of a Post-Soviet, Postmodern Latvian Identity', *Norsk Geografisk Tidsskrift* 53:2&3, 121-138.

Butzer, K. (1976) *Early Hydraulic Civilization in Egypt: A Study in Cultural Ecology* (Chicago: University of Chicago Press).

Butzer, K. (1989), 'Hartshorne, Hettner, and the Nature of Geography', in *Reflections on Richard Hartshorne's the Nature of Geography*, J.N. Entrikin and S. Brunn, (eds.). (Washington, D.C.: Association of American Geographers).

Carr, E. (1998), *Wilderness by Design: Landscape Architecture and the National Park Service* (Lincoln, NE: University of Nebraska Press).

Carroll, A.K., Zedeño, M.N., and Stoffle, R.W. (2004), 'Landscapes of the Ghost Dance: A Cartography of Numic Ritual', *Journal of Archaeological Method & Theory* 11:2, 127-56.

Cassel, J.F. (1969), *Tarahumara Indians* (San Antonio, TX: Naylor Co).

Castree, N. and Braun, B. (1998), 'The Construction of Nature and the Nature of Construction: Analytical and Political Tools for Building Survivable Futures', in *Remaking Reality: Nature at the Millennium*, B. Braun and N. Castree, (eds). (London: Routledge Press).

Chang, T.C. et al. (1996), 'Urban Heritage Tourism: The Global-Local Nexus', *Annals of Tourism Research* 23:2, 284-305.

Chape, S. et al. (compilers) (2003), *2003 United Nations List of Protected Areas* (Cambridge, UK: IUCN and UNEP-WCMC).

Chappell, J.E., Jr. (1975), 'The Ecological Dimension: Russian and American Views', *Annals of the Association of American Geographers* 65:2, 144-62.

Cock, J. and Fig, D. (2000), 'From Colonial to Community Based Conservation: Environmental Justice and the National Parks of South Africa', *Society in Transition* 31:1, 22-36.

Communauté Urbaine de Strasbourg (CUS) (1944), *Plan der Stadt Strassburg* (Communauté Urbaine de Strasbourg, Strasbourg).

Communauté Urbaine de Strasbourg (CUS) (2002), "Le Jardin des Deux Rives", Centre Documentaire, Feasability Study.

Conzen, M. (ed.) (1990), *The Making of the American Landscape* (New York and London: Routledge).

Cosgrove, D. (1984), *Social Formation and Symbolic Landscape* (London: Croom Helm).

Cosgrove, D. (1985), 'Prospect, Perspective and the Evolution of the Landscape Idea', *Transactions of the Institute of British Geographers* 10:1, 45-62.

Cosgrove, D. (1989), 'Geography is Everywhere: Culture and Symbolism in Human Landscapes', in *Horizons in Human Geography*, D. Gregory and R. Walford, (eds.). (Totowa, NJ: Barnes and Noble Books).

Cosgrove, D. (1998), *Social Formation and Symbolic Landscape, Second Edition* (Madison, WI: University of Wisconsin Press).

Cosgrove, D. and Daniels, S. (1988), *The Iconography of the Landscape: Essays on the* Symbolic Representation, Design, and Use of Past Environments (Cambridge and New York: Cambridge University Press).

Cronon, W. (2003), *Changes in the Land: Indians, Colonists, and the Ecology of New England* (New York: Hill and Wang).

Crowley, D. and Reid, S.E. (2002), *Socialist Space: Sites of Everyday Life in the Eastern Bloc* (Oxford, UK and New York: Berg).

Crump, J.R. (1999), 'What Cannot be Seen and will not be Heard: The Production of Landscape in Moline, Illinois', *Ecumene* 6:3, 295-317.

Cummings, J. (1994), *Northern Mexico Handbook: The Sea of Cortez to the Gulf of Mexico* (Chico, CA: Moon Publications).

Daniels, S. (1993), *Fields of Vision: Landscape, Imagery and National Identity in England and the United States* (Princeton, NJ: Princeton University Press).

Dann, G.M.S. and Jacobsen, J.K.S. (2003), 'Tourism Smellscapes', *Tourism Geographies* 5:1, 3-25.

Dansereau, P. (1975), *Inscape and Landscape: The Human Perception of Environment.* (New York: Columbia University Press).

Dansereau, P. (1983), 'The Template and the Impact: The Chart of Man's Course and the Charter of Man's Destiny', *INTECOL Bulletin*, 7-8, 70-109.

Davey, A.G. (1998), *National System for Protected Areas* (Cambridge, UK: IUCN).

Davis, J. (2001), 'Commentary: Tourism Research and Social Theory – Expanding the Focus', *Tourism Geographies* 3:2, 125-34.

Davis, W.M. (1909), *Geographical Essays* (New York: Dover Publications).

Demeritt, D. (1994), 'Ecology, Objectivity and Critique in Writings on Nature and Human Societies', *Journal of Historical Geography* 20:1, 22.

Dent, B. (2006) *Budapest 1956: Locations of Drama* (Budapest: Európa Könyvkiadó).

Desfor, G. and Keil, R. (2004), *Nature and the City: Making Environmental Policy in Toronto and Los Angeles* (Tucson, AZ: University of Arizona Press).

Dijkink, G. (1999), 'On the European Tradition of Nationalism and its National Code', *Geography Research Forum* 19, 45-59.

Dreyfus, F.G. (1979), *Histoire de l'Alsace* (Paris: Hachette).

Duncan, J.S. (1990), *The City as Text: The Politics of Landscape Interpretation in the Kandyan Kingdom* (New York: Cambridge University Press).

Duncan, J.S. (1993), 'Landscapes of the Self/Landscapes of the Other(s): Cultural Geography 1991-92', *Progress in Human Geography* 17:2, 367-77.

Duncan, J.S. and Duncan, N.G. (1988), '(Re)reading the Landscape', *Environment and Planning D: Society and Space* 6, 117-26.

Duncan, J.S. and Duncan, N.G. (2001), 'The Aestheticization of the Politics of Landscape Preservation', *Annals of the Association of American Geographers* 9:2, 387-409.

Dwyer, O.J. (2004), 'Symbolic Accretion and Commemoration', *Social & Cultural Geography* 5:3, 419-35.

Eco, U. (1986), *Travels in Hyperreality* (San Diego: Harcourt Brace Jovanovich).

Eco, U. (1995), 'Unlimited Semeiosis and Drift: Pragmaticism vs. "*Pragmatism*"', in *Peirce and Contemporary Thought: Philosophical Inquiries*, K.L. Ketner, (ed.). (New York: Fordham University Press).

Edensor, T. (2000), 'Staging Tourism: Tourists as Performers', *Annals of Tourism Research* 27:2, 322-44.

Edwards, J.A. and Llurdes i Coit, J.C. (1996), 'Mines and Quarries: Industrial Heritage Tourism', *Annals of Tourism Research* 23:2, 341-63.

Eriksen, T.H. (1988), 'Communicating Cultural Difference and Identity: Ethnicity and Nationalism in Mauritius', Ph.D thesis, University of Oslo.

Eriksen, T.H. (1997), 'Tensions Between the Ethnic and the Post-Ethnic: Ethnicity, Change and Mixed Marriages in Mauritius', in *The Politics of Ethnic Consciousness*, C. Govers and H. Vermeulen (eds.). (London: Macmillan).

Escobar, A. (1998), 'Whose Knowledge, Whose nature? Biodiversity, Conservation, and the Political Ecology of Social Movements', *Journal of Political Ecology* 5, 53-82.

Fontana, B.L. and Schaefer, J.P. (1997), *Tarahumara: Where Night Is the Day of the Moon* (Tucson, AZ: The University of Arizona Press).

Foote, K. (2003), *Shadowed Ground: America's Landscapes of Violence and Tragedy.* (Austin, TX: University of Texas Press).

Foote, K., Tóth, A., and Árvay, A. (2000), 'Hungary After 1989: Inscribing a New Past on Place', *The Geographical Review* 90:3, 301-34.

Forest, B. and Johnson, J. (2002), 'Unraveling the Threads of History: Era Monuments and Post-Soviet National Identity in Moscow', *Annals of the Association of American Geographers* 92:3, 524-47.

Foucault, M. (1973), *The Birth of the Clinic; An Archaeology of Medical Perception* (New York: Pantheon Books).

Foucault, M. (1977), *Discipline & Punish: The Birth of the Prison* (New York: Pantheon Books).

Foucault, M. (1980), *Power/Knowledge: Selected Interviews and Other Writings, 1972-1977*, C. Gordon, (ed.). (New York: Pantheon Books).

Foucault, M. (1984), *The Foucault Reader*, P. Rabinow, (ed.). (New York: Pantheon Books).

Foucault, M. (1990), *The History of Sexuality: An Introduction, vol. I* (New York: Vintage Books).

Fowler, W.W. (1952), *The City-State of the Greeks and Romans* (London: Macmillan and Co.).

Fürsich, E. and Robins, M. (2004), 'Visiting Africa: Constructions of Nations and Identity on Travel Websites', *Journal of Asian and African Studies* 39:1/2, 133-52.

Furze, B., De Lacy, T., and Birckhead, J. (1996), *Culture, Conservation and Biodiversity: The Social Dimension of Linking Local Level Development and Conservation Through Protected Areas* (Chichester, UK: John Wiley & Sons).

Gábor, E. (2002), *Andrássy Avenue* (Budapest: City Hall).

Gaffey, S. (2004), *Signifying Place: The Semiotic Realisation of Place in Irish Product Marketing* (Aldershot, Hants, England; Burlington, VT: Ashgate).

Garrod, B. and Fyall, A. (2001), 'Heritage Tourism: A Question of Definition', *Annals of Tourism Research* 28:4, 1049-52.

GEF (Global Environment Facility). (1999), *Project Document: Establishment of a Programme for the Consolidation of the Mesoamerican Biological Corridor* (New York: United Nations Development Programme).

Geisler, C. and de Sousa, R. (2001), 'From Refuge to Refugee: The African Case', *Public Administration and Development* 21:2, 159-70.

Germic, S. (2001), *American Green: Class, Crisis, and the Deployment of Nature in Central Park, Yosemite, and Yellowstone* (Lanham, MD: Lexington).

Ginsburg, F. (2005), 'Embedded Aesthetics: Creating a Discursive Space for Indigenous Media' in *Internationalizing Cultural Studies: An Anthology*, M.A. Abbas, J.N. Erni, and W. Dissanayake (eds.). (Malden, MA: Blackwell).

Gordon-Gentil, A. (2003), 'Norbert Benoit, historien: Chaînes cryptées', in *L'express*, 8 (31 January 2003).

Gotham, K. (2002), 'Marketing Mardi Gras: Commodification, Spectacle and the Political Economy of Tourism in New Orleans', *Urban Studies* 39:10, 1735-56.

Government of Mauritius (1988), 'White Paper on Tourism' (Port Louis, Mauritius: Ministry for Tourism).

Government of Mauritius (2003), National Heritage Fund Act. <http://www.gov.mu/portal/site/nheritage>, accessed April 2007.

Government of Mauritius (2003), *Tourism Development Plan – Mauritius and Rodrigues 2000 - 2002* (London: Emerging Markets Group Report).

Graham, B. (1998a), 'Introduction', in *Modern Europe: Place, Culture, Identity*, B. Graham, (ed.). (London: Arnold).

Graham, B. (1998b), 'The Past in Europe's Present: Diversity, Identity and the Construction of Place', in *Modern Europe: Place, Culture, Identity*, B. Graham, (ed.). (London: Arnold).

Green, N. (2005), 'Who's King of the Castle? Brahmins, Sufis and the Narrative Landscape of Daulatabad', *Contemporary South Asia* 14:1, 21-37.

Groth, P. and Bressi, T. (eds.) (1997), *Understanding Ordinary Landscapes*. (New Haven, CT: Yale University Press).

Grove, R.H. (1990), 'Colonial Conservation, Ecological Hegemony and Popular Resistance: Towards a Global Synthesis', in *Imperialism and the Natural World*, J. MacKenzie, (ed.). (Manchester, UK: Manchester University Press).

Grusin, R. (1998), 'Reproducing Yosemite: Olmsted, Environmentalism, and the Nature of Aesthetic Agency', *Cultural Studies* 12:3, 332-59.

GTZ (Deutsche Gesellschaft für Technische Zusammenarbeit). (2002), *Biodiversity in German Development Cooperation, 4th edition*. (Eschborn, Germany: Deutsche Gesellschaft für Technische Zusammenarbeit (GTZ)).

Guano, E. (2002), 'Spectacles of Modernity: Transnational Imagination and Local Hegemonies in Neoliberal Buenos Aires', *Cultural Anthropology* 17:2, 181-209.

Haeckel, E. (1899, 1992 reprint), *The Riddle of the Universe*, (Buffalo, N.Y.: Prometheus Books.

Hall, S. (1995), 'New Cultures for Old', in *A Place in the World? Places, Cultures and Globalization*, D. Massey and P. Jess, (eds.). (New York: Oxford University Press and Milton Keynes, UK.: The Open University Press).

Harris, H. and Lipman, A. (1986), 'Viewpoint: A Culture of Despair: Reflections on 'Post-Modern' Architecture', *The Sociological Review* 34:4, 837-54.

Harrison, R.P. (1992), *Forests: The Shadow of Civilization* (Chicago: The University of Chicago Press).

Hart, J.F. (1998), *The Rural Landscape* (Baltimore, MD: Johns Hopkins University Press).

Harvey, D. (1979), 'Monument and Myth', *Annals of the Association of American Geographers* 69:3, 362-81.

Harvey, D. (1989), *The Condition of Postmodernity: An Enquiry into the Origins of Cultural Change* (Oxford, Eng. and Cambridge, MA: Basil Blackwell).

Harvie, C. (1994), *The Rise of Regional Europe* (London: Routledge).

Heckenberger, M.J., et al. (2003), 'Amazonia 1492: Pristine forest or cultural parkland?' *Science* 301:5640, 1710-14.

Henderson, J. (2001), 'Heritage, Identity and Tourism in Hong Kong', *International Journal of Heritage Studies* 7:3, 219-35.

Herf, J. (1997), *Divided Memory: The Nazi Past in the Two Germanys* (Cambridge, MA: Harvard University Press).

Herlihy, P.H. (1997), 'Indigenous Peoples and Biosphere Reserve Conservation in the Mosquitia Rain Forest Corridor, Honduras' in *Conservation Through Cultural Survival: Indigenous Peoples and Protected Areas*, S. Stevens, (ed.). (Washington, D.C.: The Island Press).

Heynen, N., Perkins, H.A., and Roy, P. (2006), 'The Political Ecology of Uneven Urban Green Space: The Impact of Political Economy on Race and Ethnicity in

Producing Environmental Inequality in Milwaukee', *Urban Affairs Review* 42:1, 3-25.

Hills, M. (2002), 'The Formal and Informal Management of Diversity in the Republic of Mauritius', *Social Identities* 8:2, 287-300.

Himmelweit, S. (1991), 'Mode of Production', in *A Dictionary of Marxist Thought*, T. Bottomore (ed.). (Cambridge, MA: Blackwell).

Hoelscher, S. (2003), 'Making Place, Making Race: Performances of Whiteness in the Jim Crow South', *Annals of the Association of American Geographers* 93:3, 657-86.

Hollinshead, K. (1994), 'The Unconscious Realm of Tourism', *Tourism Management* 16:1, 707-9.

Hull, R. and Revell, G. (1989), 'Issues in Sampling Landscapes for Visual Quality Assessment', *Landscape and Urban Planning* 17:4, 323-30.

Hyndman, D. (2000), 'Dominant Discourses of Power Relations and the Melanesian Other: Interpreting the Eroticized, Effeminizing Gaze', *National Geographic Cultural Analysis* 1:1, 1-23.

Illyés, Gy. (1989), 'Egy mondat a zsarnokságról', in *Rendületlenül a hazaszeretet versei*, I. Benedek, (ed.). (Officina Nova Kiadó).

Jackson, J.B. (1970), *Landscapes: Selected Writings of J.B. Jackson* (Amherst, MA: University of Massachusetts Press).

Jackson, J.B. (1984), 'Discovering the Vernacular Landscape', in *Human Geography: An Essential Anthology*, J. Agnew, D. Livingstone, and A. Rodgers, (eds.). (Oxford, Eng. and Cambridge, MA: Blackwell).

James, B. (1999), 'Fencing in the Past: Budapest's Statue Park Museum', *Media, Culture & Society* 21:3, 291-311.

Jansen, K. (1998), *Political Ecology, Mountain Agriculture, and Knowledge in Honduras* (Amsterdam: Thela Publishers).

Jansen, K. (2000), 'Structural Adjustment, Peasant Differentiation and the Environment in Central America' in *Disappearing Peasantries? Rural Labour in Africa, Asia and Latin America*, D. Bryceson, K. Cristóbal, and J. Mooij, (eds.). (London: Intermediate Technology Publications).

Jellicoe, G. and Jellicoe, S. (1975), *The Landscape of Man: Shaping the Environment from Prehistory to the Present Day* (London: Thames and Hudson).

Jones, J.R. (1990), *Colonization and Environment: Land Settlement Projects in Central America* (Tokyo: The United Nations University).

Jones, M. (1991), 'The Elusive Reality of Landscape: Concepts and Approaches in Landscape Research', *Norsk Geografisk Tidsskrift* 45:4, 229-44.

Jordan, T.G. (1970) 'The Texan Appalachia', *Annals of the Association of American Geographers* 60:3 409-27.

K. Horváth, Z. (2005), 'Harc a Szocializmusért Szimbolikus Mezőben', *Szazad Vég* 35:1, 31-68.

Kadyrbaev, M.K. (1977), *Naskal'nyeIizobrazheni'i`a khrebta Karatau* (Alma-Ata: Nauka).

Kates, R. (1962), *Hazard and Choice Perception in Flood Plain Management* (Chicago: University of Chicago, Department of Geography Research Papers, No. 78).

Katz, C. (1998), 'Whose Nature, Whose Culture? Private Productions of Space and the 'Preservation' of Nature', in *Remaking Reality: Nature at the Millennium*, B. Braun and N. Castree, (eds.). (London: Routledge Press).

Kaufmann, E., and Zimmer, O. (1998), 'In Search of the Authentic Nation: Landscape and National Identity in Canada and Switzerland', *Nations and Nationalism* 4:4, 483-510.

Kennedy, J.G. (1978), *Tarahumara of the Sierra Madre: Beer, Ecology, and Social Organization* (Arlington Heights, IL: AHM Pub. Corp.).

Kerr, J.L. and Donovan, F.P. (1968), *Destination Topolobampo; The Kansas City, Mexico & Orient Railway* (San Marino, CA: Golden West Books).

Kirtsoglou, E. and Theodossopoulos, D. (2004), 'They are taking our culture away', *Critique of Anthropology* 24:4, 135-58.

Kniffen, F. (1965) 'Folk housing: key to diffusion' *Annals of the Association of American Geographers* 55:4, 549-77.

Knight, A. (1986). *The Mexican Revolution* (Cambridge & New York: Cambridge University Press).

Knight, D. (1983), 'Identity and Territory: Geographical Perspectives on Nationalism and Regionalism', *Annals of the Association of American Geographers* 72:4, 514-31.

Kosztolányi, D. (1963), *Édes Anna* (Budapest: Szépirodalmi Könyvkiadó).

Kulturális sugárút (2006), <http://www.sugarut.com/> (home page), accessed 1 April 2006.

Ladd, B. (1997) *The Ghosts of Berlin: Confronting German History in the Urban Landscape* (Chicago: The University of Chicago Press).

Lefebvre, H., Nicholson-Smith, D. (trans.) (1991), *The Production of Space* (Oxford, UK: Blackwell).

Leib, J.I. (2002), 'Separate Times, Shared Spaces: Arthur Ashe, Monument Avenue and the Politics of Richmond, Virginia's Symbolic Landscape', *Cultural Geographies* 9:3, 286-312.

Leitner, H. and Kang, P. (1999), 'Contested Urban Landscapes of Nationalism: The Case of Taipei', *Ecumene* 6:2, 214-33.

Lewis, P. (1979), 'Axioms for Reading the Landscape', in *The Interpretation of Ordinary Landscapes,* D.W. Meinig, (ed.). (New York: Oxford University Press).

Ley, D. and Duncan, J. (1993), 'Epilogue' in J. Duncan and D. Ley, (eds.). *Place/Culture/Representation* (London: Routledge).

Light, D. (2000), 'Gazing on Communism: Heritage Tourism and Post-communist Identities in German, Hungary and Romania', *Tourism Geographies* 2:2, 157-76.

Lisle, D. (2004), 'Gazing at Ground Zero: Tourism, Voyeurism, and Spectacle', *Journal of Cultural Research* 8:1, 3-22.

Lister, C.F. and Lister, R.H. (1966), *Chihuahua: Storehouse of Storms* (Albuquerque: University of New Mexico Press).

Lorzing, H. (2001), *The Nature of Landscape: A Personal Quest* (Rotterdam: 010 Publishers).

Lowenthal, D. (1961), 'Geography, Experience and Imagination: Towards a Geographical Epistemology,' *Annals of the Association of American Geographers* 51:3, 241-60.

Lowenthal, D. (1972), 'Geography, Experience, and Imagination; Towards a Geographical Epistemology', in *Man, Space, and Environment: Concepts in Contemporary Human Geography*, P. English and R. Mayfield, (eds.). (New York: Oxford University Press).

Lowenthal, D. and Prince, H.C. (1965), *English Landscape Tastes* (New York: American Geographical Society).

Lowenthal, D. and Prince, H.C. (1972), 'English Landscape Tastes', in *Man, Space and Environment*, P. English and R. Mayfield, (eds.). (New York: Oxford University Press).

Lutz, C. and Collins, J. (1993), *Reading National Geographic* (Chicago, IL: University of Chicago Press).

Ly Thio Fane Pineo, H. (1984), *Lured Away: The Life History of Indian Cane Workers in Mauritius* (Moka, Mauritius: Mahatma Gandhi Institute).

Lymer, K. (2001), 'Shimmering Visions', *Expedition* 46:1, 16-21.

Lymer, K. (2004), 'Rags and Rock Art: The Landscapes of Holy Site Pilgrimage in the Republic of Kazakhstan', *World Archaeology* 36:1, 158-72.

MacCannell, D. (1976), *The Tourist: A New Theory of the Leisure Class* (New York: Schocken Books).

MacCannell, D. (1986), 'Keeping Symbolic Interaction Safe from Semiotics: A Response to Harmon', *Symbolic Interaction* 9:1, 161-8.

MacCannell, D. (2001), 'Tourist Agency', *Tourism Studies* 1:1, 23-37.

Mackenney, R. (1989), *The City-State, 1500-1700: Republican Liberty in an Age of Princely Power* (Atlantic Highlands, NJ: Humanities Press International, Inc.).

MacLachlan, C. and Beezley, W. (1994), *El Gran Pueblo: A History of Greater Mexico* (Englewood Cliffs, NJ: Prentice Hall).

Mahatma Gandhi Institute (MGI) (2003), 'Aapravasi Ghat – The Landing Place of Immigrants', (Moka, Mauritius). Unpublished manuscript.

Maksimova, A.G. (1985), *Naskal´nye izobrazheni´i`a urochishcha Tamgaly*. (Alma-Ata : '·Oner').

Mar´´i`ashev, A.N. (1994), *Petroglyphs of South Kazakhstan and Semirechye* (Almaty: A.H. Margulan Archaeology Institute: Pilgrim Firm; Zaman Co).

Mathewson, K. (1998), 'Cultural Landscapes and Ecology, 1995–96: Of Oecumenics and Nature(s)', *Progress in Human Geography* 22:1. 115-28.

Mathewson, K. (2000), 'Cultural Landscapes and Ecology III: Foraging/Farming, Food, Festivities', *Progress in Human Geography* 24:3 457-74.

Mathieson, A. and Wall G. (1982), *Tourism: Economic, Physical and Social Impacts.* (London & New York: Longman).

Marshall, D. (2002), 'The Problem of the Picturesque', *Eighteenth-Century Studies* 35:3 413-37.

McDowell, L. (1983) 'Towards an Understanding of the Gender Division of Urban Space', *Environment and Planning D: Society and Space* 1:1, 59-72.

McKercher, B. and Du Cros, H. (2002), *Cultural Tourism: The Partnership Between Tourism and Cultural Heritage Management* (New York: The Hawthorn Hospitality Press).

McNeely, J. (1990), 'The Future of National Parks', *Environment* 32:1, 16-41.

Meinig, D.W. (1979), 'The Beholding Eye: Ten Versions of the Same Scene', in D.W. Meinig (ed.), *The Interpretation of Ordinary Landscapes: Geographical Essays* (New York: Oxford University Press).

Mejia, D.A. (1993), *Informe del Sexto Viaje al Parque Nacional de Celaque, Serie Miscelanea de CONSEFORTH* 24-6:93, (Siguatepeque, Honduras: Proyecto Conservación y Mejoramiento de los Recursos Forestales de Honduras).

Michalski, S. (1998), *Public Monuments: Art in Political Bondage 1870-1997* (London: Reaktion Books).

Midtgard, M. (2003), 'Authenticity – Tourist Experiences in the Norwegian Periphery' in *New Directions in Rural Tourism*, D. Hall, L. Roberts and M. Mitchell (eds.). (Aldershot, Hants., Eng. and Burlington, VT: Ashgate Publishing Company).

Miller, R. (1983), 'The Hoover® in the Garden: Middle-Class Women and Suburbanization , 1850-1920', *Environment and Planning D: Society and Space* 1:1, 73-88.

Mills, C. (1988), 'Life on the Upslope: The Postmodern Landscape of Gentrification', *Environment and Planning D, Society & Space* 6:21, 69-89.

Minca, C. and Oakes, T. (eds.) (2006), *Travels in Paradox: Remapping Tourism.* (London: Rowman & Littlefield).

Mitchell, D. (1994), 'Landscape and Surplus Value: The Making of the Ordinary in Brentwood, CA', *Environment and Planning D: Society and Space* 12:1, 7-30.

Mitchell, D. (2001), 'The Lure of the Local: Landscape Studies at the End of a Troubled Century', *Progress in Human Geography* 25:2, 269-81.

Mitchell, D. (2003), 'Cultural Landscapes: Just Landscapes or Landscapes of Justice?' *Progress in Human Geography* 27:6, 787-96.

Mitchell, W.T. (2002) *Landscape and Power, Second Edition* (Chicago: University of Chicago Press).

Mitscherlich, A. and Mitscherlich, M. (1975), *The Inability to Mourn: Principles of Collective Behavior* (New York: Grove Press).

Monk, J. (1992), 'Gender in the Landscape: Expressions of Power and Meaning', in *Inventing Places: Studies in Cultural Geography*, K. Anderson, (ed.). (Sydney: Longman Chestvie).

Morgan, W.B. and Moss, R.P. (1965), 'Geography and Ecology: The Concept of the Community and its Relationship to Environment', *Annals of the Association of American Geographers* 55:2, 339-50.

Nash, D. (1996), *The Anthropology of Tourism* (Tarrytown, NY: Elsevier Science).

Naveh, Z. and Lieberman, A. (1994), *Landscape Ecology: Theory and Application* (New York: Springer-Verlag).

Nelson, H.J. (1959), 'The Spread of Artificial Landscape Cover Over Southern California', *Annals of the Association of American Geographers* 49:3/2, 80-99.

Nelson, J.G. and Byrne, A.R. (1966), 'Man as an Instrument of Landscape Change: Fires, Floods and National Parks in the Bow Valley, Alberta', *The Geographical Review* 56:2, 226-38.

Neumann, R. (1998), *Imposing Wilderness: Struggles over Livelihood and Nature Preservation in Africa* (Berkeley: University of California Press).

Nietschmann, B.Q. (1973) *Between Land and Water: The Subsistence Ecology of the Miskito Indians, Eastern Nicaragua* (New York: Seminar Press).

Nikolaevich, A.; Bajpakov, K.M., (ed.).; Alexandrova, S.V. and Braden, K. (trans.) (1994), *Petroglyphs of South Kazakhstan and Semirechye* (Almaty: Institute of Archaeology of the National Academy of the Republic of Kazakhstan).

Nogue, J. and Vicente, J. (2004), 'Landscape and National Identity in Catalonia', *Political Geography* 23, 113-32.

Nohlen, K. (1984), 'Strasbourg au temps de l'annexion', *Monuments Historique* No. 135, Octobre-Novembre 1984, 47-53.

Nohlen, K. (1997), *Contruire une capitale Strasbourg imperial de 1870 a 1918: Les bâtiments officiels de la Place Impérial* (Collection 'Recherches et documents', tome 56, publications de la societe savante d'Alsace).

Nuryanti, W. (1996), 'Heritage and Postmodern Tourism', *Annals of Tourism Research* 23:2, 249-60.

Olcott, M.B. (1995), *The Kazakhs* (Stanford, CA: Sanford University Press, Hoover Institution Press).

Olwig, K.R. (1996), 'Recovering the Substantive Nature of Landscapes', *Annals of the American Association of Geographers* 86:4, 630-53.

Olwig, K.R. (2002), *Landscape, Nature, and the Body Politic: From Britain's Renaissance to America's New World* (Madison: University of Wisconsin Press).

Ooi, C.S. (2003), 'The Poetics and Politics of Destination Branding', Unpublished paper.

Orser, C.E. (2006), 'Symbolic Violence and Landscape Pedagogy: An Illustration from the Irish Countryside', *Historical Archaeology* 40:2, 28-44.

Osborne, B.S. (1998), 'Constructing Landscapes of Power: The George Etienne Cartier Monument, Montreal', *Journal of Historical Geography* 24:4, 431-58.

Osorio Vargas, J. (2002), 'Hacia Una Agenda Del Desarrollo Sustentable En Chile', in *El Periplo Sustentable* (Universidad Autónoma del Estado de México).

Oviedo, M.I. (1999), *Informe – Diagnostico Sobre la Situacion de las Familias Reubicada en la Zona de Amortiguamiento del Parque Nacional Celaque* (Santa Rosa de Copan, Honduras: Proyecto Celaque).

Paasi, A. (2001), 'Europe as a Social Process and Discourse', *European Urban and Regional Studies* 8:1, 7-28.

Palmer, C. (1999), 'Tourism and the Symbols of Identity', *Tourism Management* 20:3, 13-321.

Patano, S. and Sandberg, L.A. (2005), 'Winning Back More than Words? Power, discourse and quarrying on the Niagara Escarpment', *Canadian Geographer-Geographe Canadien* 49:1, 25-41.

Patin, T. (1999), 'Exhibitions and Empire: National Parks and the Performance of Manifest Destiny', *Journal of American Culture* 22:1, 41-59.

Peleggi, M. (1996), 'National Heritage and Global Tourism in Thailand', *Annals of Tourism Research* 23:2, 432-48.

Pellow, D.N. and Brulle, R.J. (eds.) (2005), *Power, Justice, and the Environment: A Critical Appraisal of the Environmental Justice Movement* (Cambridge, MA: MIT Press).

Penning-Rowsell, E. and Lowenthal, D. (1986), *Landscape Meanings and Values* (London: Allen and Unwin).

Perez, J.C. (1997), 'Barrancas Del Cobre Tourism Project Questioned by Area Ngo's', in *Border Lines*: Americaspolicy.org.

Perkins, H.C. and Thorns, D.C. (2001), 'Gazing or Performing?' *International Sociology* 16:2, 185-205.

Pinder, D. (2000), '"Old Paris is no More:" Geographies of Spectacle and Anti-Spectacle', *Antipode* 32:4, 357-86.

Pipkin, J.S. (2003), 'Glances from the Shore: Thoreau and the Material Landscape of Cape Cod', *Journal of Cultural Geography* 20:2, 1-19.

Pletsch, C. (1993), 'Regimes of Nature', *Humanist* 53:6, 3-9.

Poria, Y., Butler, R., and Airey, D. (2001) 'Clarifying Heritage Tourism', *Annals of Tourism Research* 28:4, 1047-49.

Portillo, N.P. (1997), *Geografía de Honduras* (Tegucigalpa, Honduras: Colonia Miraflores).

Poulter, G. (2004) 'Montreal and Its Environs: Imagining a National Landscape, 1867-1885.' *Journal of Canadian Studies* 38:3, 69-93.

Prohászka, L. (2004), *Szoborhistóriák* (Budapest: Városháza).

Puczkó, L. and Rátz, T. (2006), 'Managing an Urban World Heritage Site: The Development of the Cultural Avenue Project in Budapest', in *Managing World Heritage Sites*, A. Leask and A. Fyall (eds.). (Oxford: Butterworth-Heinemann).

Raat, W.D. and Janeček, G. (1996), *Mexico's Sierra Tarahumara: A Photohistory of the People of the Edge* (Norman, OK: University of Oklahoma Press).

Raup, H.F. (1959), 'Transformation of Southern California to a Cultivated Land', *Annals of the Association of American Geographers* 49:3/2, 58-78.

Redford, K.H., Robinson, J.G., and Adams, W.M. (2006), 'Parks as Shibboleths', *Conservation Biology* 20:1, 1-2.

Richards, G. (1996), 'Production and Consumption of European Cultural Tourism', *Annals of Tourism Research* 23:2, 261-83.

Ringer, G. (ed.) (1998), *Destinations: Cultural Landscapes of Tourism* (London: Routledge).

Rojek, C. and Urry, J. (eds.) (1997), *Touring Cultures: Transformations of Travel and Theory* (London and New York: Routledge).

Romsics, I. (1999), *Hungary in the Twentieth Century* (Budapest: Corvina Osiris).

Rose, G. (1995), 'Place and Identity: A Sense of Place', in *A Place in the World? Places, Cultures and Globalization*, D. Massey and P. Jess (eds.). (New York: Oxford University Press and Milton Keynes, Eng.: The Open University).

Rosenfeld, G. (2000), *Munich and Memory: Architecture, Monuments and the Legacy of the Third Reich* (Berkley, CA: University of California Press).

Rowntree, L. (1996), 'Cultural Landscape Concept in American Human Geography', in *Concepts in Human Geography*, C. Earle et al. (eds.). (Lanham, MD: Rowman and Littlefield).

Rowntree, L. and Conkey, M. (1980), 'Symbolism and the Cultural Landscape', *Annals of the Association of American Geographers* 70:4, 459-74.

Rozwadowski, A. and Pietrzak, A. (trans.) (2004), *Symbols through Time* (Poznan: Institute of Eastern Studies, Adam Mieckiewicz University).

Ryan, C. (1991), *Recreational Tourism: A Social Science Perspective* (London: Routledge).

Saarinen, T. (1966), *Perception of Drought Hazard on the Great Plains* (Chicago: University of Chicago, Department of Geography Research Papers, Number 106).

Sakellariou, M.B. (1989), *The Polis-State Definition and Origin* (Paris: Diffusion de Boccard).

Sauer, C.O. (1925), 'The Morphology of Landscape', in *Land and Life: A Selection from the Writings of Carl Ortwin Sauer*, J. Leighly (ed.). (Berkeley: University of California Press).

Sauer, C.O. (1941), 'Foreword to Historical Geography', *Annals of the Association of American Geographers* 31:1, 1-24.

Sauer, C.O. (1956), 'The Agency of Man on the Earth' In *Man's Role in Changing the Face of the Earth*, W. L. Thomas (ed.). (Chicago: University of Chicago Press).

Schein, R. (1997), 'The Place of Landscape: A Conceptual Framework for the American Scene', *Annals of the Association of American Geographers* 87, 660–80.

Schmidt, M. (2006), Personal Interview (Budapest: 13 July 2006).

Schmidt, M. (ed.), Major, A. (trans.) (2003), *Terror Háza Andrássy út 60. House of Terror* (Budapest: Public Endowment for Research in Central and East-European History and Society).

Schönle, A. (2000), 'Gogol, the Picturesque, and the Desire for the People: A Reading of "Rome"', *Russian Review* 59:4, 597-613.

Schulenburg, A.H. (2003). '"Island of the Blessed": Eden, Arcadia and the Picturesque in the Textualizing of St. Helena', *Journal of Historical Geography* 29:4, 535-53.

Secretaria de Agricultura, Ganaderia, Desarrollo Rural y Pesca (SAGARPA) 'Análisis De Los Principales Cultivos Establecidos En El Estado De Chihuahua 1995-2000.' Gobierno del Estado de Chihuahua, Mexico, 2001.

Sessions, G. (1991), 'Ecocentrism and the Anthropocentric Detour', *ReVision* 13:3, 109-16.

Semple, E.C. (1911), *Influences of Geographic Environment, on the Basis of Ratzel's System of Anthropo-geography* (New York: Holt and Co.).

Shafer, C.L. (1990), *Island Theory and Conservation Practice* (Washington, D.C.: Smithsonian Institution Press).

Simmons, I.G. (1993), *Environmental History: A Concise Introduction* (Oxford: Blackwell).

Slater, E. (1993), 'Contested Terrain: Differing Interpretations of County Wicklow's Landscape', *Irish Journal of Sociology* 3:1, 23-55.

Sletto, B. (2002), 'Producing Space(s), Representing Landscapes: Maps and Resource Conflicts in Trinidad', *Cultural Geographies* 9:4, 389-420.

Smethurst, D. (2000), 'Mountain Geography', *The Geographical Review* 90:1, 35-56.

Smith, A. (1991), *National Identity* (London: University of Nevada Press).

Smith, J. (1993), 'The Lie that Blinds: Destabilizing the Text of Landscape', in *Place/ Culture/Representation*, J. Duncan and D. Ley, (eds.). (London: Routledge).

Soper, A., Knudsen, D.C., and Yespembetova, A.S. (2003), 'Gazing at Almaty: Tourism, Meaning and the Intersection of Gazes', Paper Presented at the 99th Annual Meetings of the Association of American Geographers, New Orleans, Louisiana, March 5-8, 2003.

Sopher, D.E. (1979), 'The Landscape of Home: Myth, Experience, Social Meaning' in *The Interpretation of Ordinary Landscapes: Geographical Essays*, D.W. Meinig, (ed.). (New York: Oxford University Press).

Sørensen, S. (1997), *The County of Viborg: Cultural and Natural History Guide* (Copenhagen: Museums Council of Viborg and Høst & Søn).

Southworth, J., Nagendra, H., and Tucker, C. (2002), 'Fragmentation of a Landscape: Incorporating Landscape Metrics into Satellite Analyses of Land-Cover Change', *Landscape Research* 27:3, 253-69.

Southworth, J., Tucker, C., and Munroe, D. (2002), 'The Dynamics of Land-Cover Change in Western Honduras: Exploring Spatial and Temporal Complexity', *Agricultural Economics* 27:3, 355-69.

Squire, S. (1994), 'Accounting for Cultural Meanings: The Interface Between Geography and Tourism Studies Re-examined', *Progress in Human Geography* 18:1, 1-16.

Statue Park n.d., a. *Statue Park: Gigantic Memorials From the Communist Dictatorship* (Budapest: Ákos Réthly).

Statue Park n.d., b. *Statue Park Brochure* (Budapest: Statue Park).

Stevens, S. (1997), 'The Legacy of Yellowstone', in *Conservation through Cultural Survival: Indigenous Peoples and Protected Areas*, S. Stevens, (ed.) (Washington, DC: The Island Press).

Stilgoe, J.R. (1982), *Common Landscape of America, 1580 to 1845* (New Haven: Yale University Press).

Stoddart, D.R. (1965), 'Geography and the Ecological Approach: The Ecosystem as a Geographic Principle and Method', *Geography* 50, 242-51.

Teelock, V. (1999), "Historic' Preservation: Whose History?' in *Education and Culture at the Dawn of the Third Millennium*, I. Asargally, (ed.). (Vacoa, Mauritius: Editions Le Printemps).

Terror Háza n.d. *Terror Háza Andrássy út 60* (House of Terror Brochure) (Budapest: Terror Háza).

Till, K. (2005), *The New Berlin: Memory, Politics, Place* (Minneapolis: University of Minnesota Press).

Timothy, D. and Boyd, S. (2003), *Heritage Tourism* (Essex, England: Pearson Education Limited).

Townsend, D. (1997), 'The Picturesque', *The Journal of Aesthetics and Art Criticism* 55:4, 365-76.

Trade and Environment Database (TED) (1996), 'TED Case Studies, Volume 5, Number 2, June, 1996' (Washington, D.C.: American University, School of International Service).

Tuan, Y.F. (1968), 'Discrepancies Between Environmental Attitude and Behaviour: Examples from Europe and China', *Canadian Geographer* 12:3, 176-91.

Tuan, Y.F. (1971), *Man and Nature*, Association of American Geographers Resource Paper #10 (Washington, D.C.: Association of American Geographers).

Tuan, Y.F. (1972), 'Topophilia: Personal Encounters with the Landscape', in *Man, Space and Environment*, P.W. English and R.C. Mayfield, (eds.). (Oxford: Oxford University Press).

Tuan, Y.F. (1974a), 'Space and Place: Humanistic Perspective', Progress in Human Geography' *Progress in Geography* 6, 233-246.

Tuan, Y.F. (1974b), *Topophilia: A Study of Environmental Perception, Attitudes, and Values* (Englewood Cliffs, NJ: Prentice-Hall, Inc).

Tuan, Y.F. (2003), 'Perceptual and Cultural Geography: A Commentary' *Annals of the Association of American Geographers* 93:4, 878-81.

Tunbridge, J. and Ashworth, G.J. (1996), *Dissonant Heritage: The Management of the Past as a Resource in Conflict* (Chichester: Wiley).

Turner, B. (2002), 'Contested Identities: Human-Environment Geography', *Annals of the Association of American Geographers* 92:1, 52-74.

Turner, M., Gardner, R., and O'Neill, R. (2001), *Landscape Ecology in Theory and Practice: Pattern and Process* (New York: Springer).

United Nations Educational, Scientific and Cultural Organization (UNESCO) (2005), 'Aapravasi Ghat (Mauritius) Application for World Heritage Site Status No. 1227', <http://whc.unesco.org/en/list/1227/documents>, accessed April 2007.

United Nations Educational, Scientific and Cultural Organization (UNESCO) (2006), 'Petroglyphs within the Archaeological Landscape of Tamgaly', <http://whc.unesco.org/en/list/1145>, accessed October 18, 2006.

Unruh, J.D., Heynen, N.C., and Hossler, P. (2003), "The Political Ecology of Recovery from Armed Conflict: The Case of Landmines in Mozambique", *Political Geography* 22:8, 841-61.

Urry, J. (1990), *The Tourist Gaze: Leisure and Travel in Contemporary Society* (London: Sage).

Urry, J. (1992a), 'The Tourist Gaze and the Environment', *Theory and Culture* 9:1, 1-26.

Urry, J. (1992b). 'The Tourist Gaze Revisited', *American Behavioral Scientist* 36:2, 172-87.

Urry, J. (1995), *Consuming Places* (London and New York: Routledge).

Urry, J. (2002), *The Tourist Gaze* (Sage Publications: London). Second Edition.

Uslucan, H. (2004), 'Charles Peirce and the Semiotic Foundation of Self and Reason', *Mind, Culture, and Activity* 11:2, 96-108.

van Eeden, J. (2004), 'The Colonial Gaze: Imperialism, Myths, and South African Popular Culture', *Design Issues* 20:2, 18-33.

Vassberg, L.M. (1993), *Alsatian Acts of Identity: Language Use and Language Attitudes in Alsace* (Clevedon, Eng. and Philadelphia: Multilingual Matters Ltd.).

Wampler, J. (1969), *New Rails to Old Towns: The Region and Story of the Ferrocarriles Chihuahua Al Pacífico* (Berkeley: J. Wampler).

West, R.C. (1958), 'The Lenca Indians of Honduras: A Study in Ethnogeography', in *Latin American Geography: Historical-Geographical Essays, 1941-1998*, R.C. West, (ed.). (Baton Rouge, LA: Geoscience Publications).

Whelan, Y. (2002), 'The Construction and Destruction of a Colonial Landscape: Monuments to British Monarchs in Dublin Before and after Independence', *Journal of Historical Geography* 28:4, 508-33.

White, G. (1945), *Human Adjustments to Floods: A Geographical Approach to the Flood Problem in the United States* (Chicago, IL: University of Chicago Press).

White, H. (1973), *Metahistory: The Historical Imagination in Nineteenth-Century Europe* (Balitomre: Johns Hopkins University Press).

Wilcken, N. (2000), 'Strasbourg et l'architecture publique dans le Reichsland (1871-1918)', in *Strasbourg 1900 naissance d'une capitale,* M. Liénart, (ed.). (Strasbourg: Musées de Strasbourg, Somogy Édition d'art).

Wilson, L.D. and McCranie, J.R. (2004), 'The Herpetofauna of the Cloud Forests of Honduras', *Amphibian Reptile Conservation* 3:1, 34-48.

World Conservation Monitoring Centre (WCMC) (1998), *Workshop Report: Tropical Montane Cloud Forests Planning and Advisory Workshop, July 1998* (Cambridge, UK: World Conservation Monitoring Centre).

Wright, R.G. and Mattson, D.J. (1996), 'The Origin and Purpose of National Parks and Protected Areas', in *National Parks and Protected Areas: Their Role in Environmental Protection*, R.G. Wright, (ed.). (Cambridge, MA: Blackwell Science).

Yeoh, B.S.A. and Huang, S. (1998), 'Negotiating Public Space: Strategies and Styles of Migrant Female Domestic Workers in Singapore', *Urban Studies* 35:3, 583-602.

Yespembetova, A. (2005), *Tourism Development in the Republic of Kazakhstan*, Master's thesis. Indiana University.

Yin, R.K. (2003) *Case Study Research: Design and Methods* (Thousand Oaks, CA.: Sage Publications).

Zaring, J. (1977), 'The Romantic Face of Wales', *Annals of the Association of American Geographers* 67:3, 397-418.

Zimmerer, K.S. (1996), 'Ecology as Cornerstone and Chimera in Human Geography', in C. Earle et al. (eds.), *Concepts in Human Geography* (Lantham, MD: Rowman and Littlefield).

Zimmerer, K.S. (2000), 'The Reworking of Conservation Geographies: Nonequilibrium Landscapes and Nature-Society Hybrids', *Annals of the Association of American Geographers* 90:2, 356-69.

Zingg, R.M., et al. (2001), *Behind the Mexican Mountains* (Austin: University of Texas Press).

Zukin, S. (1991), *Landscapes of Power: From Detroit to Disney World* (Berkeley: University of California Press).

Index

Page numbers in italics refer to illustrations.

Tamgaly landscape and, 122–24
see also heritage tourism; nostalgia
Hoelsher, S., 17
Honduras, *see* Celaque National Park
Hong Kong, 55–6
Horváth, K., 93
House of Terror (Budapest), 74, 76–9, 81–7, 92–4
Huang, S., 17
human agency perspective, 9–10, 12, 13–14, 105
human rights, 48
humanistic geography, 101–2
Hungary, 74–94

iconic landscape types, 15–16, 119
identity
    defined, 20–21
    individuation process of, 34–5
    landscape and, 1, 20, *21*, 21–2
    landscape-identity-tourism nexus, 6–7, 133
    simultaneous historical identities, 19–21, 35, 131
    see also ethnic identity; local identity; national identity; regional identity; supranational identity
ideology, 49, 76, 126–27, 132
indigenous people
    in Copper Canyon region (Mexico), 41, 44
    geography of exclusion, 97, 100–101, *104*, 104–5, *106*
    loss of indigenous languages, 99
    in Mauritius, 53
individuation, 34–5
insiders
    defined, 16–17
    agricultural areas and, 113–14
    landscape and, 101–3, 110–12
    ritual meaning in tourist sites, 124–26
    views of tourists, 118
    see also intermediaries; local identity; outsiders
intermediaries, 4, 5, 51–2; *see also* local identity; outsiders; representation; state; tour guides
irony, 91–2

Jackson, J. B., 12–13
Jardin des Deux Rives, 31–2, 35

Johnson, J., 67
Jones, M., 9

Kazakhstan, *see* Tamgaly
Kehl (Germany), 31–2
Kierkegaard, Søren, 101
Knudsen, D. C., 121

landscape
    defined, 12, 20
    cultural vs. physical landscape, 9, 134
    as geographical concept, 95
    iconic landscape types, 15–16, 119
    identity and, 1, 20, *21*, 21–2
    identity-landscape-tourism nexus, 6–7, 133
    as ideology, 126–27
    insider vs. outsider perspectives of, 101–3, 110–12
    landscape approach to tourism, 133–34
    landscape ecology, 11–12
    national landscapes, 37
    Nazi landscape tourism, 71–3
    organic vs. universal landscape, 104
    place and, 124
    as power, 16, 67
    as text, 15
    Thy sublime landscape, 115–16, 118
    tourist gaze and, 4–5, 132–33
    see also cultural landscape; place; space; tourism landscape
landscape painting, 14–15, 96, 102–3, 104
Landschaft, 95, 96–7, 104–5
language
    dialect as regional identity, 33–4
    labeling of space, 32–3
    loss of indigenous languages, 99
    national identity and, 31
    Peircian vs. Saussurean semiotics, 111
    recorded narratives as representation of experience, 83–4
Lash, Scott, 3
Lefebvre, H., 65–6, 76, 134
Leib, J. I., 17
Lenca people (Honduras), *99*, 99–100, *104*, 104–5, *106*
Lewis, Pierce, 14
Ley, D., 67
Lieberman, A., 11
Livingstone, D. N., 12
local identity

For Product Safety Concerns and Information please contact our
EU representative GPSR@taylorandfrancis.com Taylor & Francis
Verlag GmbH, Kaufingerstraße 24, 80331 München, Germany